Animating Empire

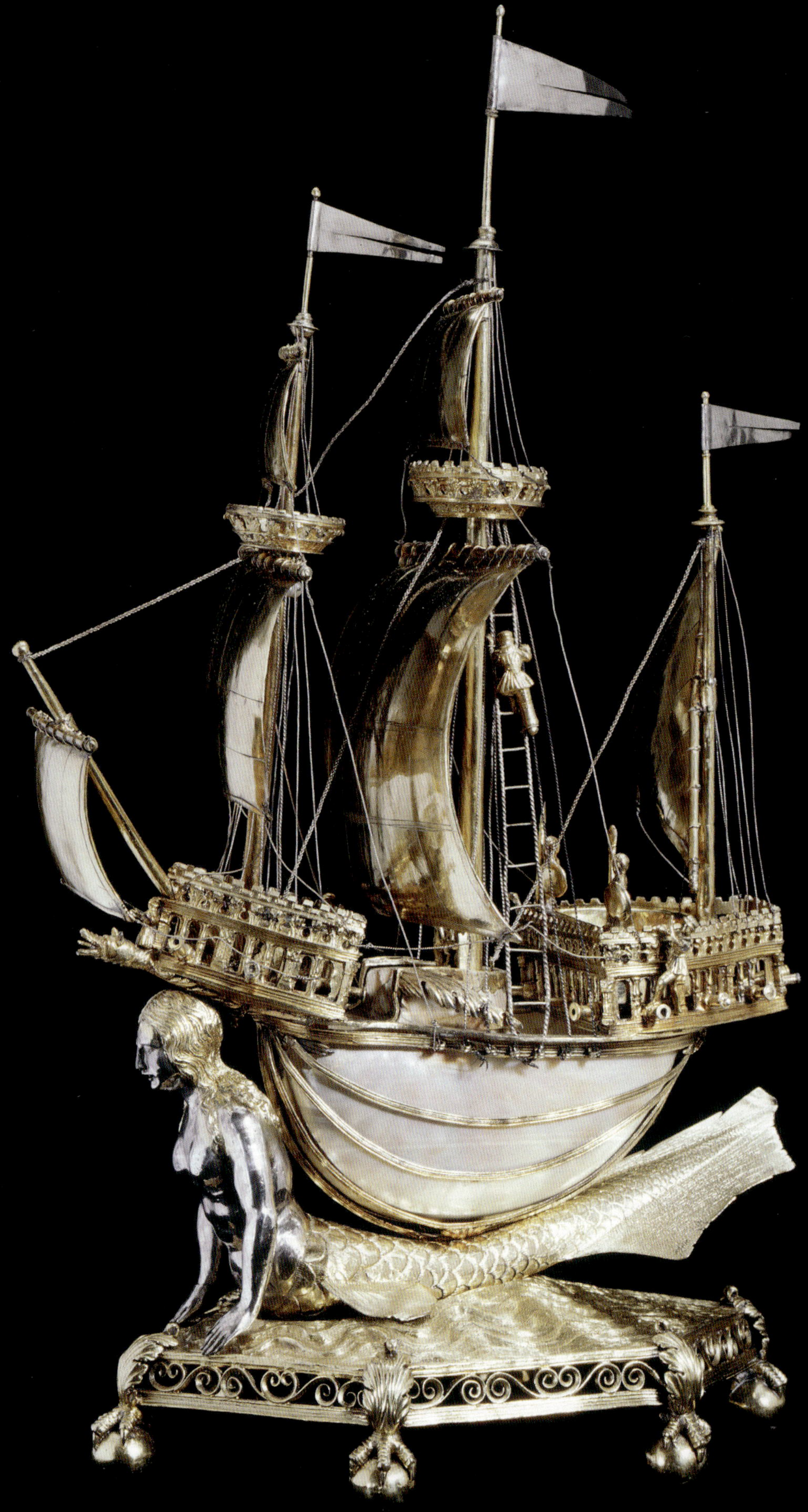

Animating Empire

Automata, the Holy Roman Empire, and the Early Modern World

Jessica Keating

The Pennsylvania State University Press
University Park, Pennsylvania

ART HISTORY
PUBLICATION INITIATIVE

This book is made possible by a collaborative grant from the Andrew W. Mellon Foundation.

Library of Congress Cataloging-in-Publication Data
Names: Keating, Jessica, 1979– author.
Title: Animating empire : automata, the Holy Roman Empire, and the early modern world / Jessica Keating.
Description: University Park, Pennsylvania : The Pennsylvania State University Press, [2018] | Includes bibliographical references and index.
Summary: "Recounts the histories of German clockwork automata, which were given as gifts and collected in the Holy Roman Empire, the Ottoman Empire, and Mughal Empire in the sixteenth and seventeenth centuries"—Provided by publisher.
Identifiers: LCCN 2017033795 | ISBN 9780271080024 (cloth : alk. paper)
Subjects: LCSH: Robots—Holy Roman Empire—History—16th century. | Robots—Holy Roman Empire—History—17th century. | Robots in art.
Classification: LCC TJ211 .K43 2018 | DDC 629.80943/09031—dc23
LC record available at https://lccn.loc.gov/2017033795

Printed in Korea.
Published by The Pennsylvania State University Press, University Park, PA 16802-1003

The Pennsylvania State University Press is a member of the Association of American University Presses.

It is the policy of The Pennsylvania State University Press to use acid-free paper. Publications on uncoated stock satisfy the minimum requirements of American National Standard for Information Sciences—Permanence of Paper for Printed Library Material, ANSI Z39.48–1992.

Frontispiece: Burghley Nef, 1527–1528, Victoria and Albert Museum, London.

Design and composition by Chris Crochetière, BW&A Books, Inc.

For my parents,

Lisa, Phil, and Eddie

Contents

Illustrations

Acknowledgments

Many institutions, colleagues, friends, and family members have contributed to the production of this book. I am profoundly grateful to five professors at Northwestern University: Claudia Swan, David van Zanten, Ann Gunter, Jesús Escobar, and Christopher Pinney. I am privileged to have their continued encouragement and support. To Claudia Swan, *meiner Doktormütter*, I owe the most profound thanks. Her knowledge, her guidance, and her friendship are precious to me, and I will be forever grateful for the boundless care and devotion she has shown me over what has now become many sweet years. While at Northwestern and after, Larry Lipking has not only been an unofficial mentor to me, but also an example of a scholar and a human being, and I thank him with love. After my time in Evanston I was extremely fortunate to be advised and taken care of by Daniela Bleichmar, Vanessa Schwartz, and Peter Mancall at the University of Southern California. Their work, camaraderie, and fortitude afforded me the confidence to write this book. I thank them for being substitute Jewish parents and for believing in me.

Numerous friends and colleagues have offered their support and brilliant ideas, and some have read and reread all or portions of this book. I am especially indebted to Lia Markey, Justin Steinberg, Christina Normore, Katie Chenoweth, Hans Wietzke, Melissa Haynes, J. K. Barrett, Stephanie Schrader, Marco Ruffini, JB Shank, Michael Gaudio, Yael Rice, Pieter Diemer, Dorothea Diemer, Prof. Dr. Iris Lauterbach, Helge Braun, and Clement Clarke. I also thank Jill Bugajski, Alison Fisher, Laura Leaper, Jacob Lewis, Lily Woodruff, Victoria Sancho-Lobis, Maureen Warren, Kristoffer Neville, Mark Rosen, Ellery Foutch, Zirwat Chowdhury, Daniel Nolan, Jason Leddington, Todd Hedrick, Laura Reagan, Sean Roberts, Alex Knodell, Rebecca Zorach, Touba Ghadessi, Adeeb Khalid, and Lyle Massey.

Research for this book was carried out in three phases, first in Germany and then back in the United States. I have enjoyed the generous support of grants and fellowships from Northwestern University and the Samuel H. Kress Foundation, from which I received the two-year History of Art Residential Fellowship at the Zentralinstitut für Kunstgeschichte. A Solmsen Postdoctoral Fellowship at the Institute for Research in the Humanities at the University of Wisconsin-Madison, and an Early Modern Studies Institute / Visual Studies Research Institute Mellon Postdoctoral Fellowship at the Huntington Library and the University of Southern California enabled me to write this book, and I am grateful to the staff, directors, and other visitors at all these institutions for making my time so fruitful. Without the cooperation of and support of colleagues at Carleton

College, I would not have been able to finish this project; I thank Katie Ryor, Baird Jarman, and Ross Elfline in particular. Terra La Chance, Heidi Eyestone, Fiona Fraser, Sara McAuliffe, and Noah Scheer have been tremendously helpful throughout the completion of this manuscript, and I thank them all warmly. From the outset of my time at Carleton, my faculty mentor, Larry Cooper, has helped sustain my work and myself. I thank him with admiration and affection.

I am deeply indebted to the staff of libraries and collections where I have worked in the United States, Germany, and Austria: the Deering Library and Special Collections at Northwestern University; the Newberry Library, Chicago; the Zentralinstitut für Kunstgeschichte, Munich; the Bayerische Staatsbibliothek, Munich; the Bayersiche Nationalmuseum, Munich; the Grünes Gewölbe, Dresden; the Kupferstichkabinett, Dresden; the Österreichische Nationalbibliothek, Vienna; the Weltliche Schatzkammer, Vienna; the Kunsthistorisches Museum, Vienna; the Museum für Angewandte Kunst, Cologne; the Huntington Library, San Marino; and the Getty Research Institute, Los Angeles. I am pleased to thank the Andrew W. Mellon Art History Publication Initiative Fund for defraying the costs of producing this book.

In the meantime I have treated some of the material of this book in a special issue of the journal *Art History*: "Metamorphosis at the Mughal Court," in "Objects in Motion in the Early Modern World," ed. Daniela Bleichmar and Meredith Martin, special issue, *Art History* 38 (2015): 732–47. It was subsequently reprinted as "Metamorphosis at the Mughal Court," in *Objects in Motion in the Early Modern World,* ed. Daniela Bleichmar and Meredith Martin (London: Wiley Blackwell, 2016), 136–51.

Ellie Goodman at Penn State University Press has been unflagging in her interest and support in this project; her expert guidance and efficiency made the publication of this book a pleasure. I must also mention the eagle eyes of Laura Reed-Morrisson, which were precious to me in the final stages of production.

A devoted group of friends and family have awaited this publication for a very long time. Mandy Gagel and Allison Hawkins continued to be my closest of friends from childhood, through college, graduate school, postdocs, and now a job. I could not have done any of it without their companionship. A heartfelt thanks to Patrice Keating, Joyce Bowersock, Carol and Rick Fine, David and Gwyneth Anne Freedman, and Lois Freedman. My sister, Becca Keating, and my stepfather, Phil Odenweller, have encouraged and supported me throughout, and I thank them dearly. My mother, Lisa Keating, has been unflinchingly by my side my entire life. Words cannot capture my profound love for her. Since September 2010 a day has not passed when I do not remember my beloved father, Eddie Keating. In many ways this book was written for him.

Northfield
February 2017

One

Early Modern Automata: An Abridged History

This book recounts the lost histories of six clockwork-powered, self-propelled, mimetic objects that were gifted, collected, and displayed during the sixteenth and seventeenth centuries in the Holy Roman Empire, the Ottoman Empire, and the Mughal Empire. The German-speaking creators, transporters, owners, and viewers of these objects designated them *Uhrwerke* (clockworks). Robert Hooke (1635–1703), the English surveyor, timekeeper, architect, and author of *Micrographia* (1665), preferred "little engines," and we now fall back on the Greek term "automata," the plural of "automaton," which means "something acting of its own will."[1] These golden, silver, enameled, and bejeweled clockwork objects set in motion a wide range of figures and events, from an emaciated Saint Jerome beating his chest, Christ being flagellated, and hares hunting dogs to Atlas bearing the weight of the world on his shoulders and peacocks fanning tails of iridescent feathers. In some cases an entire scene would be animated atop a meticulously wrought static base, while in others a single figure or group of figures would traverse a surface in a straight line, circle, or square. Their movements were driven by a collection of gears, wheels, and springs painstakingly assembled by a clockmaker, and they were typically wound up and activated by someone in a position of power. The objects were almost exclusively crafted in the city of Augsburg, and they were frequently, almost entirely, commissioned by German-speaking emperors, electors, bishops, and dukes. More than two hundred such automata are recorded in inventories and other historical sources from the sixteenth and seventeenth centuries, but unfortunately the majority were lost or dispatched to the forge during the eighteenth and nineteenth centuries.[2] Once familiar to German courts and now exceptional, automata have long occupied the margins of a traditional order of art and of the conventional exegesis of art history.[3]

Automata are distinguished by their three-dimensionality and their mechanical ability to alter and fragment the point of view of the beholder, as well as to perform during multiple and sequential moments in time. This unification of temporal process and spatial progress amounts to a mimetic mode distinct from those of other early modern works of art in two and three dimensions. But like most art objects, automata derived their significance from the imagery, composition, and material makeup of existing visual and material culture—from monumental sculpture and retable altarpieces to colossal prints, coins, jewelry, and printed broadsheets. Far from being ersatz replacements of a product of nature, as they are usually construed, early modern automata advertised—by way of their iconography and their composition—their reliance on previous representations and their elaboration of narratives only implied in their static models. With this in mind, when considering early modern automata it may be useful to reverse or invert Jean Baudrillard's oft-cited dictum about automata from *Simulacra and Simulation*. Instead of imagining that the automaton "has no other destiny than to be ceaselessly compared to a living man," it is more precise to say that "the automaton has no other destiny than to be ceaselessly compared to representations made by man."[4]

There is perhaps nothing more unnatural about the clockwork automaton than its inability to instigate its own locomotion and its dependence on the motion of a driving torque. Early modern automata had to be wound, and since they did not possess constant driving torque, they ran down and required rewinding. It was always an outside agent's desire that brought about the vivification of automata. And in the case of the sixteenth- and seventeenth-century German clockwork automata examined here, that agent was generally someone in a position of power.

Take, for instance, the Trumpeter Automaton (fig. 1) that the Wittelsbach duke Wilhelm V of Bavaria (1548–1626) gifted to the Habsburg archduke of Tyrol, Ferdinand II (1529–1595).[5] The object features two groups of five trumpeters and a single drummer, all of whom stand behind a balustrade on a dark ebony base. Cast from silver with hands and faces enameled white, and clad in gilt-silver breastplates, capes, and cylindrical caps, the trumpeters clutch their instruments with their right hands. Hanging from the ten trumpets are red wool bunting flags emblazoned with the red-white-red *Bindenschild* of Austria. The alternating colors of the Austrian shield appear again on the skins of the two drums situated before the drummer. Standing roughly twenty-five centimeters tall, the ebony base has been carved to resemble ceremonial architecture, like that constructed for the joyous entry of an early modern ruler into a city or the public festivities surrounding a noble person's rite of passage—a baptism, marriage, or funeral. In fact the whole scene has the character of a festive performance, reduced. The façade carries a centrally placed rusticated arched portal. Below the portal are two flights of stairs. Two tamed gold lions flank the prophet Daniel on the face of the stairs. Stationed symmetrically on both sides of the portal are two placards. On the left, the Bavarian coat of arms is enameled in white, blue, gold, black, and green against a gilded bronze ground, while the placard on the right displays the initials and title of the presenter of the object, "W.H.I.B." (Wilhelm Herzog in Bayern) and the year 1582, the date the automaton was completed by the jeweler Valentin Drausch (d. 1586) and the clockmaker Hans Schlottheim (1545–1626) in the Free Imperial City of Augsburg.[6] Here Wilhelm makes his inscription, with its monogram and date on a neutral zone of the base, appear as if it is a sign of authorship. But there is more, for inside the base is an iron clockwork mechanism, which, when activated, stirred the eleven musicians into action: the ten trumpeters raised and lowered their instruments to and from their mouths, while the drummer rhythmically pounded his drum. This dynamic fanfare was accompanied by a processional melody, sounded by a rack of ten pipes, also hidden inside the ebony base.

What we behold here is not simply the animated celebration of the house of Habsburg, the owners of this object since 1582, but rather something more specific: the subtle mingling of the Bavarian house of Wittelsbach and the Austrian house of Habsburg. The automaton's emphasis on this commingling is rather fitting, given that the automaton was presented to the Habsburg archduke Ferdinand II by the Wittelsbach duke Wilhelm on the occasion of a different sort of unification—Ferdinand II's second marriage, in the spring of 1582, to Anne Juliana Gonzaga (1566–1621), the daughter of Guglielmo

1. Valentin Drausch and Hans Schlottheim, Trumpeter Automaton, 1582, Kunsthistorisches Museum, Vienna.

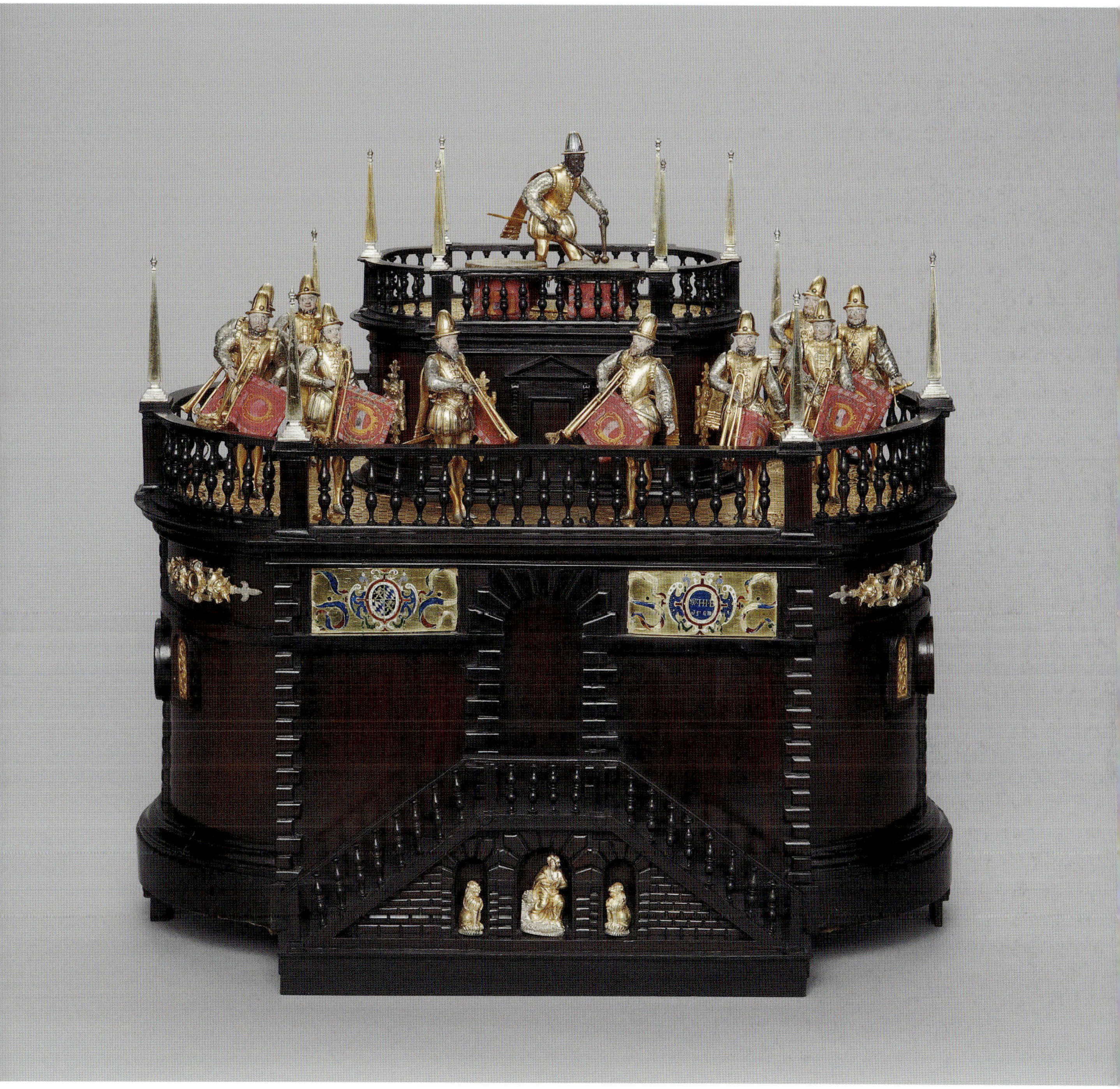

Gonzaga (1538–1587) and Eleonora of Austria (1534–1594), the duke and duchess of Mantua.[7] The Trumpeter Automaton, prominently dated 1582, tethers Ferdinand's (second) rite of passage into marriage to his relationship to Wilhelm V. Indeed, Wilhelm V had traveled from his seat of power, Munich, to Innsbruck, the locale of Ferdinand's palace in Tyrol, for the celebration of the wedding on May 14, 1582. Wilhelm was in Innsbruck not just to participate in the festivities. His mission was of far greater consequence, as he had been chosen as a proxy for Holy Roman Emperor Rudolf II (1552–1612)—Ferdinand II's nephew. Wilhelm was Rudolf's ambassador, the emperor's stand-in. This last piece of information is crucial when we consider that the bunting flags—dynamic additions that in their rustling registered the movement of the trumpeters—carry not the Tyrolean red, single-headed eagle but the Austrian shield, the ideogram of Holy Roman Emperor Rudolf II's power. These trembling flags, then, carried both physical and symbolic weight because they incorporated Rudolf II not just into the gift of the automaton but also into the wedding celebration. Moreover, the automaton attempts to make up for, or paper over, Rudolf's conspicuous absence at his uncle's wedding. In a certain respect, the trumpeter's flapping bunting flags bearing the Austrian shield are an attempt to fill the imperial void. Seen within this context, the flags read like a proleptic apology from the emperor himself. The billowing flags, set in motion by the automaton, offer to posterity the harmonious union of the Habsburgs, the Wittelsbachs, and the Gonzagas.

The purpose of this book is to reveal the political, social, and religious messages of six surviving sixteenth- and seventeenth-century German clockwork automata that were displayed or gifted between 1550 and 1625 at pivotal political moments or on the occasion of key cross-cultural encounters. The subjects of chapter 2, the London Nef and Écouen Nef (figs. 2 and 3), sizable and very complex figurations of ships in gilded silver presumably commissioned by Holy Roman Emperor Rudolf II, depict the coronation of Holy Roman Emperor Charles V (1500–1558). The Christmas Crib Automaton (fig. 14) that is the primary focus of chapter 3 is an elaborate silver automaton commissioned by Sophie of Brandenburg (1568–1622) for her husband, the elector of Saxony, Christian I (1560–1591), and placed in the Dresden Kunstkammer; it portrays the Magi and an elector of the Holy Roman Empire, in the guise of a Magus, bestowing gifts on the Christ child. The *Verkehrte Welt* (world upside-down) Automaton (fig. 25), commissioned by the Bavarian duke Albrecht V (1528–1579) and displayed in the Kunstkammer at his Residenz in Munich, features a golden monkey preaching to a group of deer; it is analyzed in chapter 4. In chapter 5 we turn to the *Sultan on Horseback* (fig. 37), one of hundreds of automata commissioned by Holy Roman Emperors Ferdinand I (1503–1564), Maximilian II (1527–1576), and Rudolf II that were part of an annual tribute payment to Ottoman sultans from 1547 to 1593; this automaton depicts an Ottoman sultan in ceremonial procession. Finally, chapter 6 is concerned with the Diana Automaton (fig. 45), a type produced in quantity in the Holy Roman Empire during the seventeenth century, and its appearance in a painting created at the court of the Mughal emperor Jahangir (1569–1627).

Automata of the sort examined in this book constitute a special class of objects precisely because of the exegetical task required for their understanding, one not just internal to their capability of movement but extending to general knowledge of the political milieu in which they were commissioned and displayed. All the objects were exhibited at courts and were meant for an audience familiar with the concerns of the particular court in which they were stored and activated. To locate and understand these objects, I turn to the minutiae that historians of art, court culture, and political history routinely wade through when they conduct their research:

inventories, court calendars, court registers, curricula of education at courts, account books, letters, patronage records, and diplomatic communiqués and gifts. But I also look beyond these sources and consult promises of benefactions made to the church, sermons, theological tracts, imperial regalia, liturgical vestments, folktales, popular broadsheets, records of courtly feasts, plays performed at court, petitions for patents addressed to the emperor, memoirs and accounts of merchants, trade routes, and the objects, images, and texts employed in missionary work done in far-flung locales. Together these objects, images, performances, and historical documents provide a richer and more textured sense of early modern automata.

While this locating process does, in and of itself, constitute valuable factual information for historians of early modern automata, in consulting these sources we also learn that clockwork automata were commissioned, displayed, and gifted during pivotal and tense political and religious moments at German-speaking courts and that their imagery and movement often relayed specific political and religious arguments. Consequently, this book also offers entry into interconnected sixteenth- and seventeenth-century worlds of great historical interest and importance—the worlds of the unmaking of a medieval empire, of the Reformation, of the Counter-Reformation, of Ottoman encroachment, and of expanded global trade. Beyond these interwoven histories, this book treats a range of themes that converge in the interpretations of the objects under discussion. The importance of technology, gift giving, courtly ceremony, and religious ceremony recurs throughout. Accordingly, the six automata examined here touch on how the social, political, and religious order of the Holy Roman Empire was represented, mechanized, and displayed. These unique and precious objects help us see how the Holy Roman Empire imagined itself during the second half of the sixteenth century and the first quarter of the seventeenth.

Automata: A Tradition

Tangled up in the court culture of the Holy Roman Empire in the second half of the sixteenth century, the German clockwork automaton emerged from a tradition stretching back to the ancient world of crafting self-propelled mimetic objects. However, as we will see, the distance between the objects examined here and their ancient and medieval precedents is surprisingly large. Within the realm of Greek poetry and myth, the animation of inanimate matter in mimetic form was closely associated with the divine forging skills of the lame, "bandy legged" god Hephaestus. Listen to Nonnus of Panopolis, the fourth-century Greek poet, describe the chariot of Hephaestus's chthonic sons, the Kabeiroi: the Kabeiroi "rode in a car of adamant; a pair of colts beat the dust with rattling hooves of brass, and they sent out a dry whinnying from their throats. These father Hephaistos had made with his *inimitable art*, breathing defiant fire between their teeth, like the pair of brazenfoot bulls which he made for Aietes the redoubtable ruler of the Colchians, with hot collars and burning pole. Eurymedon drove and guided the fiery mouths of the ironfoot steeds with a fiery bridle" (my emphasis).[8] Vivid in its evocation of the breath of life that flowed from Hephaestus to his creation, Nonnus's description of the divine forger's inimitable art was meant for a highly literate and educated audience. Writing in the fourth or fifth century of our era, Nonnus modeled his twenty-thousand-line epic of Dionysus's journey to and from India after Homer's epic works of Odysseus's Trojan expedition and homecoming. Nonnus's *Dionysiaca* looks to Homer not only in its dactylic hexameter but also in the powers and skills ascribed to the gods. Hephaestus's ability to transform metal with fire into something almost human or animal is implied in book 18 of Homer's *Iliad*:

> That done, [Hephaestus] sponged himself,
> his face, both arms, bull-neck and hairy chest,

put on a tunic, took a weighty staff,
and limped out of his workshop. Round their lord
came fluttering maids of gold, like living girls:
intelligences, voices, power of motion
these maids have, and skills learnt from immortals.
Now they came rustling to support their lord.[9]

Together Nonnus and Homer endeavor to highlight Hephaestus's divine ability not just to shape and transmute metal with fire but to animate inanimate matter, in such a way that his creations seem enlivened—ensouled. This conflation of technology and divine power is also present in ancient discussions of Daedalus, who, according to some accounts, was a mortal descendent of Hephaestus. Daedalus, an Athenian, was considered the first sculptor of the gods. But his sculptures, like Hephaestus's bronze and gold creations, had the "power of motion." Mischievous in their demeanor, Daedalus's sculptures would cause havoc, scampering away before cult rituals could be performed, playing jokes on the unwitting Athenians. In this capacity Daedalus's ensouled works are mentioned by Aeschylus, Socrates, and Aristophanes.[10] Aristotle himself, in discussing the nature of the soul in *De Anima*, wondered how Daedalus was able to enliven his creations, and proposed that he filled his hollow statues with quicksilver.[11]

The Greco-Roman world developed celebratory attitudes toward Hephaestus's fluttering maidens, Daedalus's naughty statues, and other mythological automata.[12] The immortals' divine ability to conjoin the disparate—metal, wood, self-propelled movement—both unsettled and excited. Greek and Roman engineers opted to create their own versions, objects that were propelled not by an unknowable divine force but by their own ingenuity. Ctesibius (285–222 BCE), purportedly the first keeper of the Museum of Alexandria and a pioneer in the field of pneumatics, was the initiator of the tradition of pneumatic—wind- and water-powered—automaton making.[13] Of special interest are his designs for a hydraulic self-playing organ and a clepsydra, or water clock. Although his third-century works have survived only as fragments in book 10 of Vitruvius's *De Architectura*, there is a consensus among scholars that later engineers, such as Philo Byzantium (280–220 BCE) and Hero of Alexandria (ca. 10–70 CE), relied heavily on Ctesibus's work.[14] In addition to designing fortresses and catapults, both Philo and Hero engineered complex "automatic theaters." Composed of multiple figures that interacted with each other and with their contrived surroundings, these three-dimensional moving mechanical scenes were intended for a large audience. For instance, in his *Pneumatica* Hero of Alexandria contrived a device in which a figure of Hercules, with a taut bow in hand, shoots an arrow aimed at an apple. Once the arrow was released and aloft, a pulley lifted the apple, causing Hercules to miss his target and instead skim the head of a serpent that was wrapped around the branch of a tree. As if startled, the serpent would exude a loud hiss.[15]

If it were not for numerous Arabic translations of both Philo and Hero, their work would have been lost. The thirteenth-century *Book of Knowledge of Ingenious Mechanical Devices*, penned by al-Jazarī (1136–1206) at the court of the Artuqid Sultan of Diyarbakır on the Tigris, was the most widely copied and thus influential Arabic work to relate Hero.[16] Al-Jazarī's work was dependent on an earlier ninth-century treatise also known as the *Book of Ingenious Devices*, by three brothers known as the Banu Musa. The Banu Musa's *Book of Ingenious Devices* did not, however, just preserve Hero's *Pneumatica* but also utilized the basic mechanics of Hero's automatic theaters—the pulley, the winch, the lever, water pressure, and steam—to cultivate new devices with more figures performing a greater variety of tasks.[17] Almost life-size, these pneumatically driven scenes were typically incorporated into time-telling devices. Take, for instance, "A Water Clock from which can be told the Passage of the Solar Hours." Here, "a house" twice the size of a man contained one large door with twelve smaller doors situated above it, and still twelve more doors above them—twenty-five doors in total. Between the large door and the twenty-four smaller ones extended a decorative frieze. Every hour a crescent moon would move from right to left along the edge of the frieze. As the crescent moon passed before the double-paneled doors, a figure would emerge while the corresponding single-paneled doors would flip over and change color. Meanwhile, two falcons perched in niches flanking the central door would stretch out their wings, tilt forward, and drop a ball into a vase that contained a cymbal—resounding in a crash. Every three hours two drummers beat their drums while two trumpeters blew their horns, and a figure crashed cymbals.[18]

It is tempting to regard this elaborate water clock, with its drummer and trumpeters, as a distant relative of the Trumpeter Automaton that Wilhelm of Bavaria gave to Archduke Ferdinand II. Or, to put it slightly differently, we might be tempted to consider the German clockwork automaton as a belated expression of Greek, Roman, and Islamic traditions of pneumatic automaton making. Yet I am convinced that even if the Trumpeter Automaton recalls certain aspects of al-Jazarī's ingenious device, a gulf separates the two. It is not just that early modern automata, as we will see, proclaim themselves technological objects of a particular era: often their imagery also directly engages with the political and religious issues facing the courts in which they were made, or collected, or displayed, or gifted.

A famous example of an early modern automaton that was put to political use is Leonardo da Vinci's legendary mechanical lion. According to a scrap of paper bound into the fifth volume of the unfinished *History of Florence* by Piero Parenti:

> When the King entered Milan, besides the other entertainments, Lionardo [*sic*] da Vinci, the famous painter and our Florentine, devised the following intervention: he represented a lion above the gate, which, lying down, got onto its feet when the King came in, and with its paw opened up its chest and pulled out blue balls full of gold lilies, which he threw and strewed about on the ground. Afterwards he pulled out his heart and, pressing it, more gold lilies came out, showing how the Florentine *Marzocco*, represented by such an animal, had his guts full of lilies. Stopping beside this spectacle, [the King] liked it and took much pleasure in it.[19]

Here, according to Jill Burke, Parenti is referring to the French king Louis XII (1462–1515) and his ceremonial entrance into Milan in July 1509. This wondrous lion was capable of rising to its feet, opening its chest with its own paws, disemboweling itself, and casting about lilies—the French royal flower.[20] Several other notable authors mention different occasions on which Leonardo devised an automaton of a lion for a French king. Vasari, in both the 1550 and 1568 editions of the *Vite*, relates: "During his lifetime the king of France came to Milan, and he begged Leonardo to make something unusual, and so Leonardo made a lion which walked a few steps before its chest opened, revealing it to be filled with lilies."[21] In his *Libro dei Sogni* (1564) Giovanni Paolo Lomazzo relates Vasari's account almost verbatim. But in his *Tratto dell'Arte della Pittura* (1584), Lomazzo adds that the lion was made for the French king Francis I.[22] In 1600 Michelangelo Buonarroti the Younger wrote that a lion filled with lilies was activated at the wedding of Marie de Médicis, and that this lion was "similar to that which Leonardo da Vinci realized for the Florentine nation on the occasion of Francis I's entry into Milan."[23]

Pinpointing exactly the when, where, and why of Leonardo's famed lion interests me less than the fact that animated beasts were, on all accounts, made for nobility and as a result were associated with rulership. The link between self-propelled mimetic objects and rulership has a deep history, one that stretches back before al-Jazarī dedicated his *Book of Knowledge of Ingenious Mechanical Devices* to the Artuqid Sultan of Diyarbakır in the thirteenth century. Roaring lions, singing birds, and alighting eagles, were, according to Jewish legend, integrated into King Solomon's throne at his palace in Jerusalem. Only Solomon's presence on the seat of the throne could activate the automata. Other later accounts report that the throne could also rotate 360 degrees, as well as take flight. This terrestrial throne was considered by some Jewish authors to be an earthly manifestation of God's heavenly throne—the seat of the cosmocrat.[24]

Solomon's throne was later imitated by Byzantine emperors. Although now lost, the throne lives on in a lengthy description by Liudprand of Cremona (920–972), an Italian bishop who traveled to Constantinople as a papal envoy in 949:[25]

> Before the emperor's seat stood a tree made of bronze gilded over, whose branches were filled with birds, also made of gilded bronze, which uttered different cries, each according to its varying species. . . . The throne was of immense size and was guarded by lions, made of either bronze or of wood covered over in gold, who beat the ground with their tails and gave a dreadful roar with open mouth and quivering tongue. Leaning upon the shoulders of two eunuchs, I was brought into the emperor's presence. At my approach the lions began to roar and the birds cry out. . . . After I had three times made obeisance to the emperor with my face upon the ground, I lifted my head and behold! The man, whom just before I had seen sitting on a moderately elevated seat had now changed his raiment and was sitting on the level of the ceiling. How it was done I could not imagine, unless perhaps he was lifted up by some such sort of device as we use for raising the timbers of a wine press.[26]

As recalled by Liudprand of Cremona, the imperial throne was teeming with signs of animal life. The animals do not just move and emit roars and song but also appear to perceive and react to the approaching and ritually submissive imperial audience. There is no mention of the highly choreographed and ritualized elements of an imperial reception, which would allow for the design and performance of such theater. It is as if, at least in Liudprand's rhetoric, the show had not been programmed or rehearsed—as if the performance of the Byzantine throne of Solomon were devoid of human intervention. Yet alongside the intensity with which the description dwells on the lifelike animal and avian sounds, movements, and behaviors, and their elicitation of dread in the supplicant, the description of the throne calls attention to its artificial and manufactured aspects. For instance, Liudprand is keen to identify the materials out of which the throne is constructed—bronze, wood, and gold. Moreover, we know that most if not all the animals that seem to move themselves, as if alive, are golden. This automaton is marked by the dissonance between the lifelike animation of the representation and the way the representation calls attention to its artificiality through its materiality. However much the throne of Solomon attempts to represent the power of the emperor through its wondrous animation, and however much it impressed its intended audience of foreigners, the Byzantine throne remains a large, theatrical, and in the end conventional mode of imperial self-representation.

On all scholarly accounts, the Byzantine throne of Solomon relied on the Hellenistic tradition of pneumatic automaton construction as it came down through a variety of Greek, Latin, and Arabic manuscripts.[27] Aesthetically, these water- and air-powered devices are fundamentally different from the clockwork automata that began appearing in Europe in the fourteenth century. Born out of experimentation with metal wheels, gears, springs, and screws, the first clockwork automata were connected to large clocks, clocks that were incorporated into the architectural fabric of churches, monasteries, and civic structures.[28] Constructed between 1352 and 1354, the clock at the Strasbourg Cathedral stands as the quintessential example of this nascent tradition.[29] Situated at the south transept of the cathedral, the clock exemplified none of the concerns or ambitions of the pneumatic and hydraulic automata discussed thus far. In the iconography of the animated figures (the three Magi bowing before a figure of the Virgin, and a rooster that flapped its wings and stuck out its tongue, hailing the new dawn brought on by Christ's birth) stood a biblical and allegorical directive for understanding the clock's other animated elements—a calendar and an astrolabe that indicated the movement of the sun, moon, and planets. The ensemble's character as a public work must have been apparent to its original viewers, as it is for us today. Far from subtle, the large figures, which could be seen far below on the ground, attempted to convey a universal and clear message for all who passed by. The Strasbourg Cathedral clock maintained its significance for some time. When the clock needed refurbishment in the sixteenth century, two brothers, Isaac and Josias Habrecht, were hired to update its mechanical innards. The conservation began in 1540, was completed in 1574, and then was commemorated in Conrad Dasypodius's publication *Warhafftige Ausslegung und Beschrybung des Astronomischen Uhrwercks zu Strassbourg* (1580).[30]

Numerous other clocks with moving figures followed on the heels of the clock at Strasbourg. In 1450 in Bologna Master Giovanni Evangelista da Piacenza and Master Bartolomeo di Gnudolo designed and built a clock on the Palazzo Comunale. Like the Strasbourg clock, Bologna's featured a procession of the Magi, who presented gifts to the Virgin and child. This central act of prestation was supplemented by trumpeting angels, saints, and holy dignitaries of the church. The clock also contained a Pythagorean model of the universe—a globe composed of flames around which orbited the sun, the moon, the earth, and all the planets—designed by Cardinal Basilios Bessarion (1403–1472).[31] The city of Ghent in Flanders took a different tack. In 1510 the clock on the city's belfry was fitted with a figure of Adam, who struck the hours, and a figure of Eve, who struck the half-hours. While Adam and Eve sounded the segments of the day, a snake slithered around, tempting the original couple away from their hourly labor.[32] Even more iconographically novel was the cathedral clock built in 1514 in Munich. Here, every hour on the hour, according to Joseph Gallmayr in the *Münchner Intelligenzblatt* (1779), God the Father, wielding a sword, would emerge from behind a door while figures of the Virgin and Christ begged for his mercy on behalf of humanity.[33]

The automata examined in this book emerged unquestionably from these earlier large-scale clocks with moving figures—not just in terms of their mechanical insides, but also, as we will see, because in several cases their imagery was prefigured by them. Yet even though these public works inspired the objects discussed in the following chapters, there are significant differences that often get papered over in the literature on early modern automata. The difference in scale between the early and later automata cannot be overemphasized. Whereas the public clockwork automata were intended to be seen from a great distance in a public space, potentially by a large number of viewers from all rungs of the social ladder, the German clockwork automata commissioned, collected, gifted, and displayed at courts inside and outside the Holy Roman Empire were manufactured to be viewed in comparatively intimate settings—such as Kunstkammern and reception halls—by an exclusively elite courtly audience. Another major difference lies in the process of activation. The figures on the Strasbourg Cathedral clock, the clock in the Palazzo Comunale in Bologna, the clock on the belfry in Ghent, and the Frauenkirche in Munich were programmed to be activated at certain times of day, consistently. And the figures on these clocks performed whether or not an audience was present. Since the majority of smaller clockwork automata required manual winding to be set in motion, their performances were far less frequent, and their performances often coincided with a tense political moment in the life of the court in which they were housed. The gears, wheels, and springs only turned and clicked when an agent external to the object desired to set them in motion.

Clockmakers

But who were the artisans who painstakingly fitted all the elements of the automata together? The makers of these objects were not trained engineers like Giovanni Fontana (ca. 1395–ca. 1455), who crafted an automaton of a devil to undermine the superstitions of magically enlivened statues recounted repeatedly by theologians and Magi.[34] Nor were the artisans of German clockwork automata engineers who insisted on the importance of the mechanical arts, or engineers who used self-propelled mimetic objects to help illustrate the licit rather than illicit qualities of wonder and curiosity.[35] By and large we know very little about the clockmakers who toiled away in the dark crafting automata. The exception is Juanelo Turriano (ca. 1500–1585), an Italian-born clockmaker who worked at the court of King Philip II of Spain (1527–1598). Turriano's name is closely associated with an automaton of a Franciscan monk, now in the Smithsonian Institution. Before 1976, when the 1607–11 inventory of Holy Roman Emperor Rudolf II's Kunstkammer in Prague was discovered, Turriano was believed to have been the maker of the Cittern Player automaton in the Kunsthistorisches Museum in Vienna.[36] This assumption was in part due to an account found in Ambrosio de Morales's *Las antigüedades de las ciudades España* (1575), which reads:

> Juanelo as a diversion . . . wanted to create anew ancient statues which moved and on that account were called automata by the Greeks. He made a lady more than a tereia high who placed on a table dances all over it to the sound of a drum which she meanwhile beats her self, and goes round in circles, returning to where she started. Though it is a toy and fit for mirth, it is nevertheless a great proof of his high intelligence. Additionally, Turriano is fabled to have constructed an automaton that would, daily, traverse the streets of Toledo, marching to the Archbishop's residence. Everyday the automaton is said to have left the residence laden with an allowance of bread and meat, after doing obeisance to the donor.[37]

Whether Turriano crafted any of these objects will perhaps remain a mystery. It is worth noting, however, that Christoph Magraff, a clockmaker at the imperial court in Prague, is listed as the maker of the *Cittern Player* in the 1607–11 inventory of Rudolf II's Kunstkammer.[38]

Many if not most of the clockmakers who crafted automata collected by German-speaking courts were members of clockmakers' guilds. The earliest clockmakers were trained as blacksmiths, locksmiths, or gun founders—artisans adept at working with metal in its liquid and solid states, as well as possessing a keen ability to think spatially and logically. Especially for locksmiths and gun founders, knowing how to create interworking parts of different shapes and sizes that could provide a chain of causal and intentional links was essential to the construction of the mechanisms. From the fourteenth century onwards, sophisticated clockmaking was found in the German-speaking world. In 1307 Charles V, king of France, brought a German to his palace at Saint-Pol to construct a clock, and a hundred years later built a small clock for the French queen.[39]

Before the sixteenth century, clocks were, by and large, produced where they were consumed. Centers of clock making were yet to emerge, and it was clockmakers, not clocks, who traveled. Augsburg was one of the first and most important centers of clock production and artisanal training, owing in part to the strong tradition of metalworking in the city. In 1392 a clock was mentioned for the first time in city records: the account book of Augsburg's master builders mentions the construction of a clock on the Perlach Tower—a twelfth-century tower incorporated into the Rathaus. It was not until 1442, however, that the guild of clockmakers was established as a self-governing entity. Until that time clockmakers, their training, and their product, were regulated by the smiths' guild, the regulatory body that oversaw not just professions associated with metalworking but painting and saddlery as well. Thus 1442 marks the moment when Augsburg became a vigorous site of clock production. From that point onward, the clockmakers' guild was highly organized and fostered the training of young apprentices and journey men, in addition to the establishment of master workshops. By 1555 Augsburg contained a total 150 locksmiths, gun founders, and clockmakers, and between 1550 and 1650 fifty to seventy clock shops flourished, each staffed by at least one master, one journeyman, and one apprentice. A typical shop produced fifteen to twenty-five clocks a year. Starting in the mid-sixteenth century, most of these workshops produced spring-driven portable domestic clocks.[40] In these objects, "timekeeping is governed by an oscillator or escapement which controls the unidirectional movement of the motive power and transforms it into a slow, steady, and regular motion whose meaning appears on the face of the clock."[41] Initiating this chain reaction required clockmakers to be precise in cutting the teeth of the gears, and adept at stacking thin gear trains. The clock, or an automaton, would be a success only if each of its components' movements triggered the next movement in a predetermined pattern or program. For clocks, each movement was visible on the clock face; for automata, the movement of the internal mechanism resulted in the movement of objects or figures.

Automata and the Holy Roman Empire

It is no wonder, given that the craft of clockmaking flourished in its realm, that the Holy Roman Empire connected itself with this highly skilled form of technology. And the marriage of the empire with clockmaking and automata occurs early on in the life of this technology. For instance, in 1361 Holy Roman Emperor Charles IV (1316–1378) inaugurated a clock in the Frauenkirche in Nuremberg to commemorate the Golden Bull of 1356.[42] Each day, when the clock struck noon, the seven electors of the Holy Roman Empire would emerge and bow before a seated figure of the emperor. We will see a similar animated ritual of submission on the London and Écouen Nefs in chapter 2. The two automata, which were intended to traverse a table in a Kunstkammer or during a feast, reanimated the coronation of Holy Roman Emperor Charles V (1500–1558) almost thirty years after his death. Yet despite the striking similarities between the clock inaugurated by Charles IV and the automata carrying a representation of his namesake, the meaning and function of these animated rituals of obedience vary from context to context. The clock in the Frauenkirche in Nuremberg embodies the plenitude of imperial power, the recognition of the Holy Roman Emperor by the most powerful clerical and secular rulers within his realm, as described in the Golden Bull of 1356, and the cohesion of a universal Christian monarchy.[43] In the London and Écouen Nefs, made during Holy Roman Emperor Rudolf II's reign and long after Charles's abdication and demise, the animated ceremony assumes, as we shall see, a more elegiac tone: rather than celebrate a contemporary moment, it looks backward and lauds a medieval notion of a universal Christian monarchy as it faded into the past with the success of the Reformation.

Charles V's grandfather, Maximilian I (1459–1519), has also been linked to automata. According to Peter Ramus in his *Scholarum mathematicarum libri unus et triginta* (1569) and later in John Dee's "Mathematicall Preface" to Henry Billingsley's 1570 translation of Euclid's *Elements*, the Nuremberg mathematician Johannes Müller (1436–1476), also known as Regiomontanus, presented Maximilian with an iron eagle that flew out to escort the emperor as he approached the city gates of Nuremberg on a visit to the city.[44] The eagle, whether fabled or not, was the appropriate choice for the emperor, as it was a crucial emblem of the emperor himself, appearing again and again in representations of Habsburg emperors. As part of the spectacle of the emperor's entrance into the Free Imperial City of Nuremberg, watched by hundreds if not thousands of people anxiously awaiting the emperor's arrival, the iron eagle would seem to defy all reason: iron is not a material that easily takes flight, whether in the form of an eagle or not. The reliability of Ramus's and Dee's discussions of the bird is dubious, and in my considered opinion they were no more than a fanciful attempt at the end of the sixteenth century to lionize their respective endeavors in mathematics and engineering. A sounder connection between the imperial house and an avian automaton lies in the inventory of Rudolf II's Kunstkammer (1607–11). Here we find an automaton of a peacock that squawked and fanned real peacock feathers, and that stood on a long, green table in the main chamber of the Kunstkammer.[45] The peacock, in the context of the Kunstkammer, is doubly significant. First, like the eagle, the peacock was a Habsburg *Wappentier* and, like the *Habsburger Pfau* from 1555 housed in Schloss Ambras, is typically represented with the coats of arms of all the dominions controlled by the Habsburgs.[46] Yet if the automaton of the peacock at the heart of the imperial collection invited its viewers to recall Rudolf's extensive domains, and to marvel at its fusion of nature and artifice, the object is, it must be noted, of a mixed pedigree. The peacock is a species of bird that, during the period in question, was deeply associated with the sense of sight. The deep blue "eyes" of the peacock's plumage, according to the medieval bestiary tradition, signified the ability to foresee danger.[47] Staring back at the visitors in the Kunstkammer, the peacock announces the intense engagement of the sense of sight that the Kunstkammer demands of its visitors. All this is to say that the Holy Roman Empire, the rulers of the Holy Roman Empire, and automata were allied from a very early date. The automata could stand in for the emperor and the empire. But before moving on to discuss other automata collected and displayed by German princes, a word about the makeup and nature of the Holy Roman Empire is in order.

The Empire in the Second Half of the Sixteenth Century

All the automata discussed in this book were produced within a seventy-five-year period, roughly between 1550 and 1625, a time when the Holy Roman Empire was recovering from one jolt to its system—the Reformation—and gearing up for a second—the Thirty Years' War. It was, on the one hand, a time of principalities whose interests did not always coincide and, on the other hand, a time of relative cohesion, set in motion by the threat of the Ottoman Empire on Europe's southern and eastern borders. In the sixteenth century the empire spanned present-day Germany, Austria, the Czech Republic, and Slovenia. And despite regional and linguistic differences, German and Latin were the official administrative languages. The constitution of the sixteenth-century empire had been established with the Golden Bull of 1356, mentioned earlier. Issued at the Imperial Diet of Nuremberg and Metz, the Golden Bull sought to inaugurate a unified Christian empire by creating concord between the emperor and the imperial estates—states with rulers who had rights and who voted on imperial matters. Throughout the sixteenth century these states were governed by an ecclesiastical or secular ruler. The most influential, wealthy, and powerful were the archbishops of Cologne, Mainz, and Trier as well as the prince of the County Palatine of the Rhine, Saxony, Brandenburg, and Bohemia. Together these clerical and secular heads of state were referred to as Kurfürsten, or electors. The stem *Kur* is based on the high middle German verb *kiessen*, meaning to choose.[48] And choose they did, for they were responsible for electing the emperor. Electing the emperor had two purposes. First, since in theory the election was to be unanimous, it worked to eliminate discord among the electors. Second, it weakened imperial power. Holding the various imperial estates together was the belief in a common defense and the need for the collection of taxes—the *Gemeiner Pfennig* (Common Penny).

The protestations of Martin Luther (1483–1546) against the church and the pope fostered rupture within the empire. The Diet of Augsburg of 1521, presided over by the Holy Roman Emperor Charles V, was an attempt to neutralize Luther. There Luther was required to renounce or reaffirm his views on the church, and after his refusal to recant he was accused of heresy and denounced. Swiftly thereafter the Edict of Worms was issued; it decreed that any supporter of Luther would be punished as a heretic. As a result, the Schmalkaldic League, an alliance of Lutheran princes, was formed. These princes eventually fought a war against Charles V and other Catholic princes. Now referred to as the Schmalkaldic War, a series of skirmishes ended with Charles V and the Catholic princes recognizing the legal legitimacy of the Lutheran rulers and followers of Luther. The Peace of Augsburg of 1555 provided the legal basis for the coexistence of Lutheranism and Catholicism in the Holy Roman Empire.[49] It was in this document that the voices of the confessionally split empire first spoke in concert: "In order to bring peace into the holy Empire of the Germanic Nation between Roman Imperial Majesty and the Electors, Princes and Estates: let neither his Imperial Majesty nor the Electors, Princes &c. do any violence or harm to any estate of the Empire on account of the Augsburg Confession, but let them enjoy their religious belief, liturgy, and ceremonies as well as their estates and other rights and privileges in peace."[50] This concordance assured that neither the emperor, nor a prince, nor an elector could wage war on overtly religious grounds on imperial soil. Furthermore, the Peace of Augsburg permitted the ruler of each imperial estate to choose the estate's confession: *cuius regio, eius religio*. This binding, this *religio*, of the ruler to his subjects strengthened the power of the electors while at the same time fostering even greater distinctions among Catholic and Protestant lands, causing the ideal of a universal Christian monarchy to slowly fade away. Both the Christmas Crib Automaton and the Verkehrte Welt Automaton were objects that responded to the confessional splits and tensions in the empire.

If the Reformation helped weaken the notion of a universal Christian empire, then the encroachment of the Ottoman Empire onto European soil served to keep the idea alive. Beginning in 1526 with the Battle of Mohács and in 1529 with the failed siege of Vienna, the Ottoman Empire, led by Sultan Süleyman the Magnificent (1494–1566), became the Holy Roman Empire's most formidable enemy. Defending the empire against the Ottomans, who were regularly imagined to be the antichrist by Catholics and Protestants alike, preoccupied the imperial army and the emperor's foreign policy toward both European and non-European powers. After several years of repeated skirmishes and growing anxiety over the potential loss of the Kingdom of Hungary, the Holy Roman Empire signed a truce in 1547. The conditions of the truce stipulated that the Holy Roman Empire pay the Ottoman sultan an annual tribute of forty thousand ducats; the payment was made yearly until 1593, when war broke out between the two empires.[51] Although the relationship between the Ottomans and the Habsburgs was tense, the truce fostered a good deal of exchange between the two empires, and as we will see in chapter 5, automata played a central role in the transfer of cultural goods and wealth from the German-speaking world to the Ottoman east.

The Holy Roman Empire's contact with the east during the sixteenth century extended beyond the Ottoman Empire to the Indian subcontinent. From the beginning of the sixteenth century, mercantile families from Augsburg and Nuremberg helped fund both Spanish and Portuguese commercial and colonial enterprises in and around the Indian Ocean. And it is through these mercantile and imperial connections that the German-speaking world became a major player on the global stage.[52] Frequently these commercial relationships between the Holy Roman Empire and lands in the east were cultivated through the exchange of texts, works of art, curiosities, and even automata, as will be discussed in chapter 6.

The demise of a universal Christian monarchy, the prince's right to determine the confessional bent of his territory, the Catholic princes' acerbic response to the novel and popular salvific claims of Luther, the tenuous relationship forged with the Ottomans, and the cultural commensurability between European and foreign rulers are issues grappled with by some clockwork automata commissioned by German princes and produced between 1550 and 1625 in the Holy Roman Empire. These objects were more than manifestations of wondrous and curious imagery and of technological prowess: they were also a means through which claims about the predicament of the world could be persuasively made. In automata German rulers found a representational practice that could do political and religious work.

Collecting Automata

Automata in the Holy Roman Empire were by and large displayed and stored in princely Kunstkammern, or art chambers. During the period in question Kunstkammern were clearly designated spaces in princely residences, access to which was severely limited. In some instances an entire structure was built to house the collection, as with the Kunstkammer at the ducal residence in Munich, considered the first "Renaissance" building of the Bavarian capital. In other instances structures were altered to make room for the collection, as with the Kunstkammer of Holy Roman Emperor Rudolf II, built over the imperial stable at Hradčany palace in Prague, as well as the Kunstkammer at Schloss Ambras, for which Archduke Ferdinand II renovated a medieval fortress, referred to in a document dated 1574 as the *Unterschloss*. In Dresden, seven rooms in the elector's palace were allotted to the Kunstkammer. The first two chambers were the grandest in scale and contained the majority of the roughly ten thousand objects painstakingly collected by the elector August I (1526–1586). We gain a good impression, at a textual remove, of what a princely collection encompassed from the inventory taken in 1607–11 of Holy Roman Emperor Rudolf II's Kunstkammer. The entries record objects that were rare, curious, precious, and foreign—unicorn and rhinoceros horns, griffin claws, dragon teeth, ostrich eggs, "Indian" baskets, pipe organs from Siam, ivory caskets inlaid with mother of pearl, Mixtec feather paintings, mounted and etched nautilus shells, wax paintings, pictures made from inlaid colored stones, Chinese porcelain, paintings, engravings, portraits, small bronzes, unworked gemstones, antique sculpture, astrolabes, globes, sextants, and clocks.[53] Typically, although not always, the collection was loosely organized into five groups: naturalia, artificalia, antiquities, exotica, and scientifica. Naturalia encompassed natural objects such as plant specimens, seeds, stuffed animals, and minerals. Man-made objects fell under the rubric of artificalia: ceramics, vessels, fountains, mosaics, and tools. Extra-European things, for instance feather work from the New World, ivory saltcellars from Africa, Bengali textiles, colchas, and "idols," were often categorized as exotica. Scientific instruments, compasses, astrolabes, rulers, telescopes, and globes, among other things, were categorized as scientifica. Automata were sometimes listed in inventories under the rubric of scientifica (as in the inventory of the imperial Kunstkammer in Prague in 1607–11), or else placed under no heading at all (as in the inventory of the Kunstkammer in Dresden in 1587).

The number of automata varied from collection to collection. Rudolf's collection contained at least twenty-five automata, whereas Albrecht V's Kunstkammer housed just one. With the exception of Archduke Ferdinand II's Kunstkammer at Schloss Ambras, automata were not displayed together in Kunstkammern. The Christmas Crib Automaton, the subject of chapter 3, was exhibited alongside the bones of a primordial giant, a mother-of-pearl casket from Gujarat, elaborately turned ivory, and a mirror, the frame of which was adorned with an ostentatious allegory of good government in the elector's Kunstkammer in Dresden.[54] And according to the 1598 inventory of the ducal Kunstkammer in Munich and subsequent descriptions of the collection, the Verkehrte Welt Automaton was situated on a table in the northeast corner of the chamber, just outside the *Gewölbe* (vault) that preserved the most valuable objects in the ducal collection.[55]

This book was motivated in part by a concern to introduce a more flexible understanding of these objects' roles in princely Kunstkammern than has been put forth by the art historian Horst Bredekamp, who has stressed that automata were the crowning point of a universal history that Kunstkammern embodied.[56] Even more, it is concerned with reconsidering how early modern automata are imagined and perceived. Scholars have worked tirelessly to shed light on the intellectual currents that underpinned the manufacture and collection of early modern automata. For instance, Todd Andrew Borlik argues that the early modern drive to animate inanimate matter in mimetic form was the inevitable outcome of a late medieval alchemical tradition that ascribed extraordinary powers to human art.[57] To Wendy Beth Hyman, early modern clockwork automata were produced, collected, and displayed because they were physical manifestations of the fashionable Paracelsian formulation of the "organic continuity between art and nature."[58] To Minsoo Kang, the construction and reception of early modern automata can only be understood through an analysis of the complex link in the early modern world between the "magical" or "demonic" arts and the "mechanical arts."[59] To Jessica Riskin, early modern automata were construed, culturally, theologically, and philosophically, as pseudo-life forms full of vital and even divine capacities.[60] And in a different vein, to Alexander Marr early modern automata were at once objects and rhetorical tropes employed by well-educated engineers writing for a courtly audience, to illustrate the licit rather than illicit qualities of wonder and curiosity.[61] The book before you could not have been written without this work, because scholars such as Borlik, Hyman, Kang, Riskin, and Marr have revealed how automata were received, written about, and appropriated by a variety of discourses, from poetry to philosophy.

At the heart of this scholarship is the characterization that in one way or another, early modern automata were considered invented nature: like all oxymorons, such as "silent music" or "dark light," these objects and their makers upset the ontological order. What all interpretations of early modern automata share is the desire to excavate and explain the anxieties, powerful in both premodern and modern western societies, that artifice can be harnessed for less than benevolent purposes and that humans can lose control of their own creations. These ideas and anxieties certainly inform the commission, manufacture, and display of automata in the sixteenth and seventeenth centuries. It is my contention, however, that there are other factors in play. Accordingly, this book reconsiders these objects' imagery and animation, their various political and religious meanings and functions, and the circumstances that brought the automata into being.

In the quest to circumscribe a penumbral point on the early modern ontological continuum and to delineate the Janus face of human creative power, questions have yet to be posed about the objecthood of automata—when, where, and why they were displayed, what they represented, their material composition, the human agency required to stir them into action, or the length of their performance. If we do not ask such basic questions about early modern automata, we run the risk of relegating these distinct and exceptional objects to the role of carriers of a clandestine human desire to attain godlike status through the animation of inanimate matter. An automaton in a state of suspended animation or a lousy performance by a windup clockwork automaton, a performance marked by sudden stuttering, fading, stasis, and then repetition, has yet to be seriously addressed.

This book re-concretizes early modern automata. It traces the circumstances of their commission and production; it considers their materiality; it takes into account the duration of their program; it follows their circulation from collection to collection as well as around the globe; it analyzes their subject matter in relation to contemporary visual and material culture; and it discusses the various sensitivities and failures of early clockwork technology. As a result, the conclusion I have drawn belies the spectrum of interpretations that automata have engendered, because the crux of the book is this: the animation of inanimate matter was a means to specific political ends in the early modern period, not an end in and of itself.

Two

Ever More Variations on the Imperial Theme at the Court of Rudolf II

The 1587 inventory of the Dresden Kunstkammer painstakingly lists more than ten thousand objects conspicuously displayed in the elector of Saxony's Kunstkammer.[1] Among lavishly ornamented mirrors, masterfully designed *pietra dura* tabletops, and the bones of a giant discovered outside Dresden, the chronicler lists a mechanized, gilded silver ship. Its deck contained trumpeters heralding an enthroned Holy Roman Emperor, surrounded by seven electors and heralds of empire in a circular procession. Each of the three masts is described as supporting crow's nests, in which armed sailors (*Bussknechte*) sounded out the hours with small hammers.[2] In 1593 the French traveler Jacques Esprinchard (1573–1604) encountered a similar object in a collection in Augsburg.[3] In this instance the "horologue," whose inventor Georg Roll is identified by name and which is said to have been in the works for sixteen years already, was crafted not from precious metal but from ebony. Like the ship recorded in Dresden, the object in Augsburg was propelled by a clockwork mechanism and functioned as a moving base for the seated emperor and his electoral princes, described by Esprinchard as paying him a "grand reverance." In an inventory compiled three years later, in 1596, of the possessions of Archduke Ferdinand II (1529–1595), yet another object was recorded that bears comparison to the Dresden and Augsburg pieces. This automaton was a gift from the bishop of Augsburg, Otto Truchsess von Waldburg (1514–1573), and was stored in the *Silberkammer* (silver chamber). According to the inventory, the automaton adorned the lid of a silver vessel, on which the seven electors of empire are arrayed in a circular procession before the enthroned emperor, who was placed beneath a baldachin.[4]

These automata, as described by the late sixteenth-century accounts, correspond significantly to two surviving mechanized table ships, or *nefs*, that display the coronation of Holy Roman Emperor Charles V.[5] One housed in the British Museum and believed to have been crafted by the Augsburg clockmaker Hans Schlottheim is commonly referred to as the London Nef (fig. 2). The second, also attributed to Schlottheim, is housed in the Musée national de la Renaissance in Écouen and is known as the Écouen Nef (fig. 3).[6] The connection between these two clockwork-driven galleons and the epochal expansion of the Holy Roman Empire to the lush, resourceful, and populated New World during the reign of Charles V has been explored, fruitfully.[7] This is no wonder. The numerous landfalls along the Atlantic coastline and the subsequent claiming of that verdant land by Spanish sailors under the auspices of the Habsburg crown in the first half of the sixteenth century have become a powerful paradigm for thinking through European cultural identity, European imperial expansion, and empire itself. And the sixteenth-century galleons on which Charles

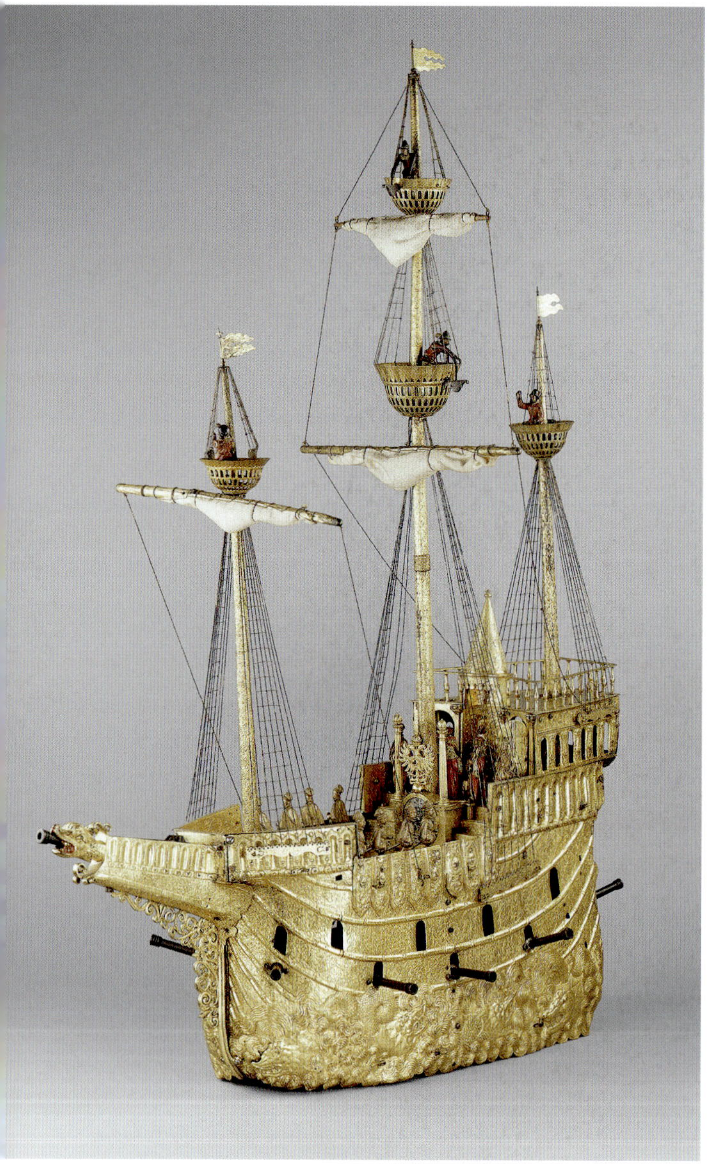

2. Hans Schlottheim, London Nef, 1580–1590, British Museum, London.

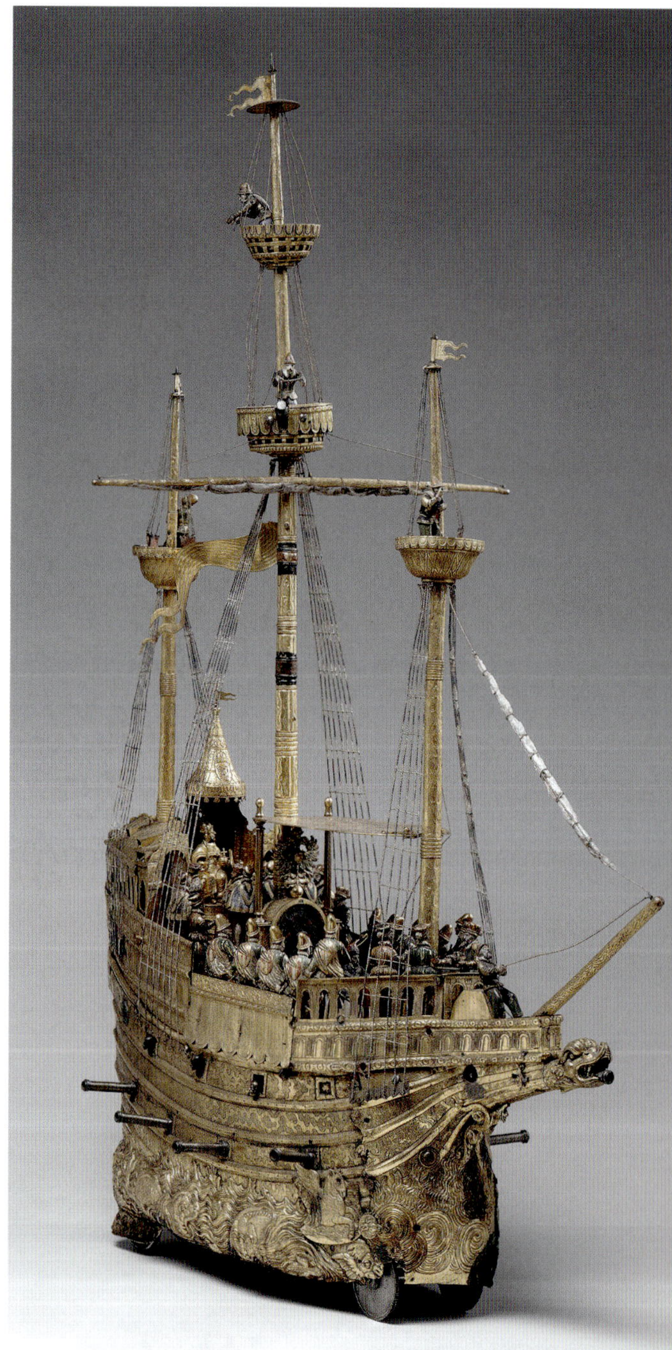

3. Hans Schlottheim, Écouen Nef, 1580–1590, Musée national de la Renaissance, Écouen.

is perched in no uncertain terms evoke this sixteenth-century enterprise. But the nefs offered their audience simultaneously yet another, obsolescent, version of empire—a vision of things past. This "phantom of empire," to borrow Frances A. Yates's phrase, haunts these objects because their conception and construction spanned a period when the nature and legitimacy of the Holy Roman Emperor and the Holy Roman Empire were increasingly questioned.[8] By rearticulating the grammar of the longevity and legitimacy of Habsburg imperial rule, the nefs' imagery and animation disclose a variation on the imperial theme in the late sixteenth century that has heretofore gone unexplored.

Predecessors

The London and Écouen Nefs are not just spectral vestiges of empire: they are also material remnants of a late medieval tradition of crafting nautical vessels to function as votive offerings, reliquaries, drinking vessels, and princely emblems. Queen Margaret of Provence (1221 1295), the wife of King Louis IX (later Saint Louis, 1214–1270), donated a small silver nef to the Church of Saint Nicolas de Port in Lorraine. According to Jean de Joinville, in 1254 the entire royal family was caught in a tempest while returning, by ship, from the Holy Land. During the storm the queen vowed that if her family were returned safely to France, she would present Saint Nicolas de Port with a nef valued at five marks of silver. The family arrived home safely and Margaret was true to her word. Sieur de Joinville tells us: "When the Queen (whom God absolve!) had returned to France she had the silver ship made in Paris. And there were in the ship the King, the Queen, and three children, all of silver; the sailors, the mast, the tillers and the ropes all of silver; and the sails all sown with silver wire. And the Queen told me the making had cost a hundred *livres*."[9] An inventory compiled on September 16, 1380, of the belongings of Charles V, king of France (1338–1330), itemizes two nefs given to the king by the city of Paris.[10] One served as a seat marker for the king at courtly feasts, while the other functioned as a saltcellar, or *saliera*. Joanna of Castile (1479–1555), mother of Holy Roman Emperor Charles V, owned a nef set on wheels that she had converted into a reliquary to house and display the relics of Santa Leocadia in Toledo.[11] Pope Gregory XI owned "a cup decorated with waves made in the form of a silver ship, gilt inside."[12]

The Burghley Nef (fig. 4), only 35 centimeters tall and 20.5 centimeters wide yet lavishly decorated, is a pristine surviving example of a nef that served as a saltcellar.[13] Like several other nefs crafted in this period, the Burghley Nef stands atop an elaborate base. Six disembodied griffin talons support the gilded silver substructure, which was carefully wrought to mimic a calm sea. Swimming atop the waves is an oversize mermaid, whose tail cradles a nautilus shell that has been overturned to resemble the hull of a ship. Several sailors populate the deck of this fantastical vessel (one even climbs the twisted silver ropes rigging the polished silver sails), while two figures play chess on the forecastle.

Nefs such as the Burghley Nef were placed on tables for major events. For instance, a nef was present at the coronation banquet for Charles VIII at Rheims on May 30, 1484.[14] An image that captures the grandeur of such a sumptuous royal event is the calendar miniature for the month of January in the Limbourg Brothers' *Très Riches Heures* (1411–12, Chantilly, Musée Condé, Ms. 65, fol. 2; fig. 5).[15] The miniature depicts the luster of a New Year's reception held at the court of Jean, Duc de Berry (1340–1460). Robed in a blue-brocaded, fur-trimmed *houppelande*, the duke sits at a banquet table, while his courtiers surround him and attend to his wishes. The trestle banquet table is crowded with vessels crafted of exceptional materials. The largest, and arguably most important, object on the table is the golden nef situated to the duke's left. The Duc de

Berry's nef stands on an elaborately ornamented hexagonal base and is crowned on the forecastles with a bear and a swan, two of the duke's heraldic animals. The animals are echoed in the red silk textile, which hangs from the chimney directly behind. In addition to calling attention to the duke's prominent position at the table, the nef appears to hold tableware, possibly the duke's personal utensils. This miniature makes clear and emphasizes the link between the ruler and the nef. In a sense, the nef functions as a literal representation of the ruler, for it would stand in his place when he was absent.

Like the nef owned by Joanna of Castile, the London and Écouen Nefs rest on wheel carriages, but instead of being pushed down the table by an external force, these objects trundled down the surface of a table or floor in a straight line by way of a clockwork mechanism. Rather than being guided by someone's hand, the London and Écouen Nefs were wound up, set down, and released. Eventually they would slow down and perhaps sputter; ultimately they came to a halt until someone set them in motion again. Another crucial distinction between the London and Écouen Nefs, on the one hand, and non-mechanical nefs, on the other, is that they were without utilitarian functions. Neither contained spaces to hold salt, spices, utensils, or personal services. Despite these differences, the London and Écouen automata pertain to a courtly tradition of commissioning nefs and displaying them at courtly feasts. Schlottheim's innovations are significant, extending to modifications in scale, imagery, and function, which together transformed these objects into vessels that made claims about the transcendental continuity of the House of Habsburg.

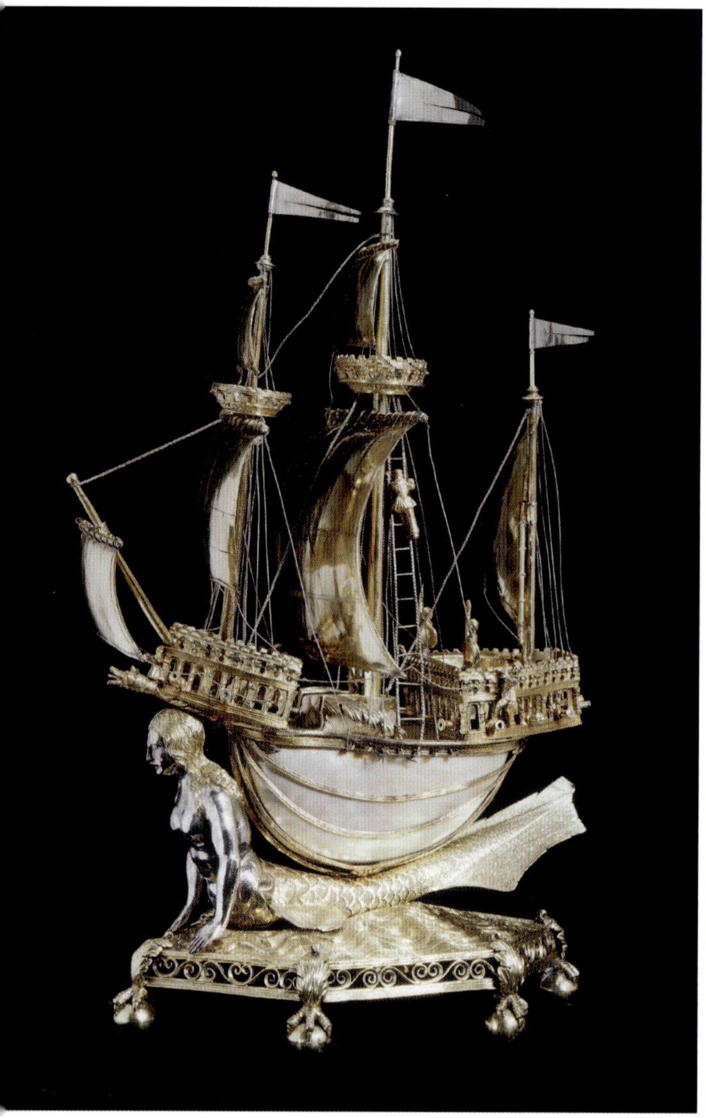

4. Burghley Nef, 1527–1528, Victoria and Albert Museum, London.

5. Limbourg Brothers, *January: New Year's Reception of the Duke Jean de Berry*, from the *Très Riches Heures*, 1411–1416, Musée Condé, Chantilly.

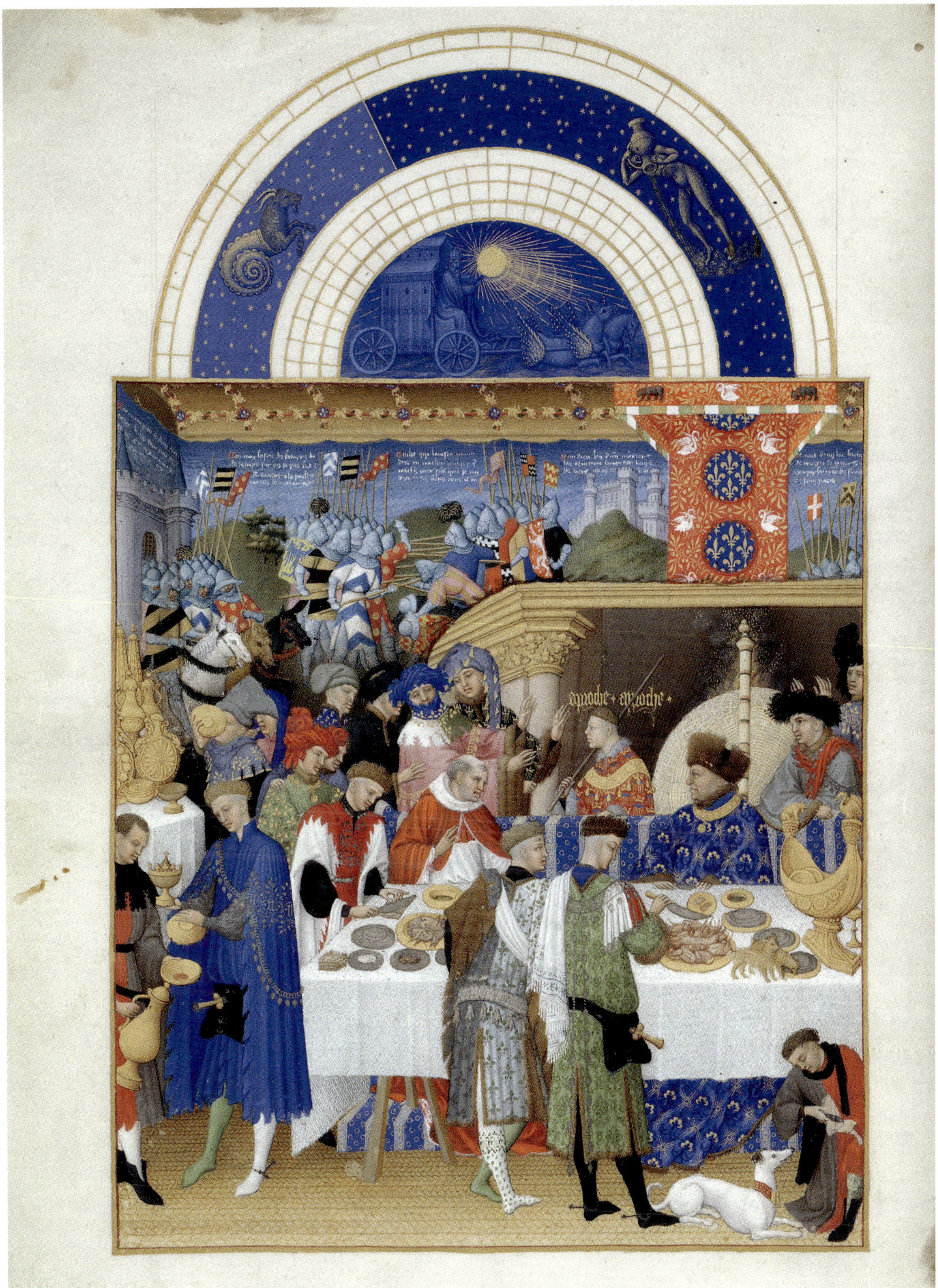
aproche · aproche ·

Hans Schlottheim and the Imperial Court in Prague

The son of a clockmaker, Hans Schlottheim was born in Naumbourg, Saxony, sometime between 1544 and 1547.[16] In his early to mid-twenties, between 1567 and 1573, he left his father's workshop and moved to Augsburg.[17] Upon his arrival he trained under the master clockmaker Jeremias Metzger, and in 1573 he married Ursula Geiger, the widow of master locksmith Hans Schitterer.[18] Schlottheim obtained his *Schmiedergerechtigkeit* (right to forge) the same year. Three years later, in 1576, he presented his *Meisterwerk* to the *geschworene Geschaumeister* (sworn inspection masters) in Augsburg and was awarded the title of *Meister*. In 1577 he installed a large clock on the façade of his house—an ingenious form of advertising.[19] Schlottheim's self-promotion paid off. He bought a second house on the famed Schmiedgasse in 1579, placing himself at the center of clockmaking in the Holy Roman Empire.[20] The absence in archival documents of any mention of Schlottheim for seven years, between 1579 and 1586, is marked. Despite the lacuna, we may deduce that the 1582 visit of Holy Roman Emperor Rudolf II to Augsburg—to convene the Diet of Augsburg—played a pivotal role in Schlottheim's career and his relations with the imperial court. As scholars have suggested, it was during Rudolf II's stay in the Free Imperial City that he probably came in contact with Schlottheim's work.[21] This is even more likely given that in 1586 Schlottheim requested and received permission from the city of Augsburg and the clockmakers' guild to leave the city and work at the imperial court in Prague, where he remained until 1589. From that year until the mid-1590s he is recorded at the Saxon court in Dresden, where he created the *Tower of Babel* (ca. 1590, Grünes Gewölbe, Dresden) and the Christmas Crib Automaton (ca. 1588, Mathematisch-Physikalischer Salon, Dresden).[22]

The dating of the London and Écouen Nefs is unstable, not only because the objects bear no date stamps but also because there exist no other comparanda for the automata. Although the objects exalt Rudolf II's deceased forebear Charles V, whose revival of medieval notions of imperial universalism Rudolf II honored, several scholars claim that the objects were made before Schlottheim entered the employ of the imperial crown.[23] Writing in 1963, Friedrich Streng claimed that the Écouen Nef was crafted in 1580 and the London Nef in 1581.[24] It is unclear on what evidence Streng based his opinion. The underlying assumption is that Schlottheim caught the emperor's attention with these two works in 1582 and was subsequently invited to Prague.[25] In point of fact it is highly improbable that Schlottheim executed the two nefs without an imperial commission or in such a short span of time. Both ships are extremely large, each measuring almost a meter in length and height, and they are made almost entirely of gilded silver. Schlottheim would have required substantial financial backing. The possibility that both nefs were speculative ventures at such an early date in the clockmaker's career is also highly unlikely. Indeed, the principal subject matter strongly suggests an imperial commission, or at the very least the backing and encouragement of someone intimate with the imperial court.

"Your throne was established long ago; you are from all eternity" (Psalm 93:2)

The technical virtuosity of the automata lies not only in the complex mechanisms that propelled them forward, housed in the hulls of the ships, but also in their highly wrought surfaces of gilded silver. The entire body of both ships is covered in finely embossed and chased *Laubwerk* or *Ranken*, a decorative motif that incorporates a variety of flora and fauna into complex arrangements of curves and counter-curves.[26] In addition to the finely detailed *Laubwerk*,

Schlottheim skillfully molded the ribbing of the vessels, which simultaneously interrupts and draws attention to the spiraling motif. Cannons emerge from inside the ships and extend outward through portals on both the port and starboard sides, while larger cannons protrude from the mouths of the dragons on both heads of the vessels.[27] Both hulls are engulfed in incised patterns of turbulent water, populated by chimeras and sea monsters swarming the vessels. On each nef, these ravaging and hostile elements pose no opposition to the migration of the imperial seat.

Brass figures in formal courtly attire are installed on the decks of the ships. The London Nef carries two groups of four male figures each, standing shoulder to shoulder on either side of the deck. More crowded, the Écouen Nef has twenty-three figures standing aboard the deck of the ship or actively working the masts. On the Écouen Nef ten trumpeters line the perimeter of the deck. A drummer, who also played, is situated between the trumpeters on both automata. The main decks each also contain a representation of a double-headed imperial eagle suspended between two columns, a significant detail that recalls Charles V's device of the two columns of Hercules.[28] Strategically placed, the imperial insigne announces and demarcates the symbolic and ritual act that unfolds beneath the conical baldachin at the back of the vessels.

Directly behind the main mast on both automata, Holy Roman Emperor Charles V sits enthroned, laden with imperial treasure and garbed in golden ecclesiastical vestments—the imperial mitre crown, topped with a cross, and a golden brocaded dalmatic, the vestment worn during the imperial unction by the pope (figs. 6 and 7).[29] On the London Nef the emperor holds the imperial scepter in his left hand, while his right is outstretched in a gesture of benediction; on the Écouen Nef the emperor holds the *Reichsapfel* (imperial orb) and wields a sword. In both cases Charles V is represented in the moment immediately following his imperial coronation, the moment immediately preceding his ontological and liturgical transformation into a *persona gemina*, "human by nature and divine by grace."[30]

The coronation scene on the London Nef incorporates the emperor's "consortium," literally his standing committee.[31] This consortium is composed of the three *Geistliche Kurfürsten* (holy electors) of the Holy Roman Empire and the four *Weltliche Kurfürsten* (secular electors), all enveloped in regal, ermine-trimmed red mantles and *Kurhüte* (electoral crowns), and arrayed in a circular procession around the newly crowned emperor (fig. 8). On the London Nef the electors are identifiable on the basis of the attributes they were required to hold in the coronation ceremony. The three Geistliche Kurfürsten carry texts. The archbishops of Mainz and Trier carry books; the archbishop of Cologne, a scroll. The four Weltliche Kurfürsten wield liturgical items used only during the coronation of the emperor. The king of Bohemia carries a silver chalice, from which the emperor drank a mixture of wine and water during the ceremony. The washing basin and cloth used to cleanse the body of the emperor in preparation for his anointment are held by the margrave of Brandenburg. The elector of Saxony carries the *Reichschwert* (imperial sword), a weapon believed to embody the power of the empire.[32] Finally, the elector of Palatine holds a large silver key to the imperial kitchen, designating his responsibility over the sovereign's food. The coronation scene on the Écouen Nef includes two additional heralds donning the imperial *Wappenrock* (tabard), who lead the procession, and an eighth elector, whose hands are empty.[33] Other differences are visible in the liturgical items that the margrave of Brandenburg and the elector of Saxony present. Instead of the washbasin and cloth, the margrave carries a scepter. This crucial piece of imperial regalia is paired with the Reichsapfel held by the elector of Saxony. As on the London Nef, the archbishops of Mainz and Trier carry books, the archbishop of Cologne a scroll, the king of

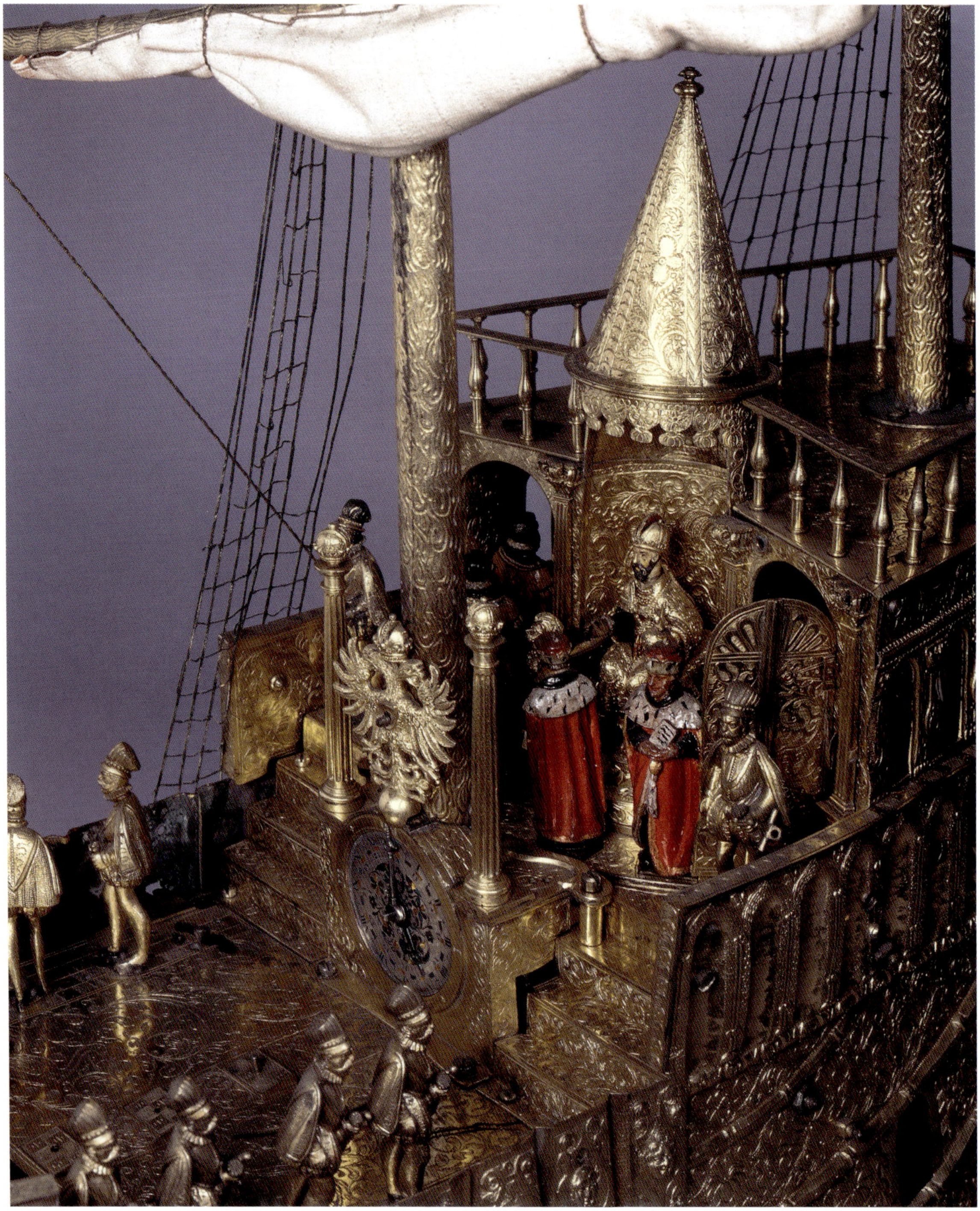

6. Hans Schlottheim, London Nef (detail), 1580–1590, British Museum, London.

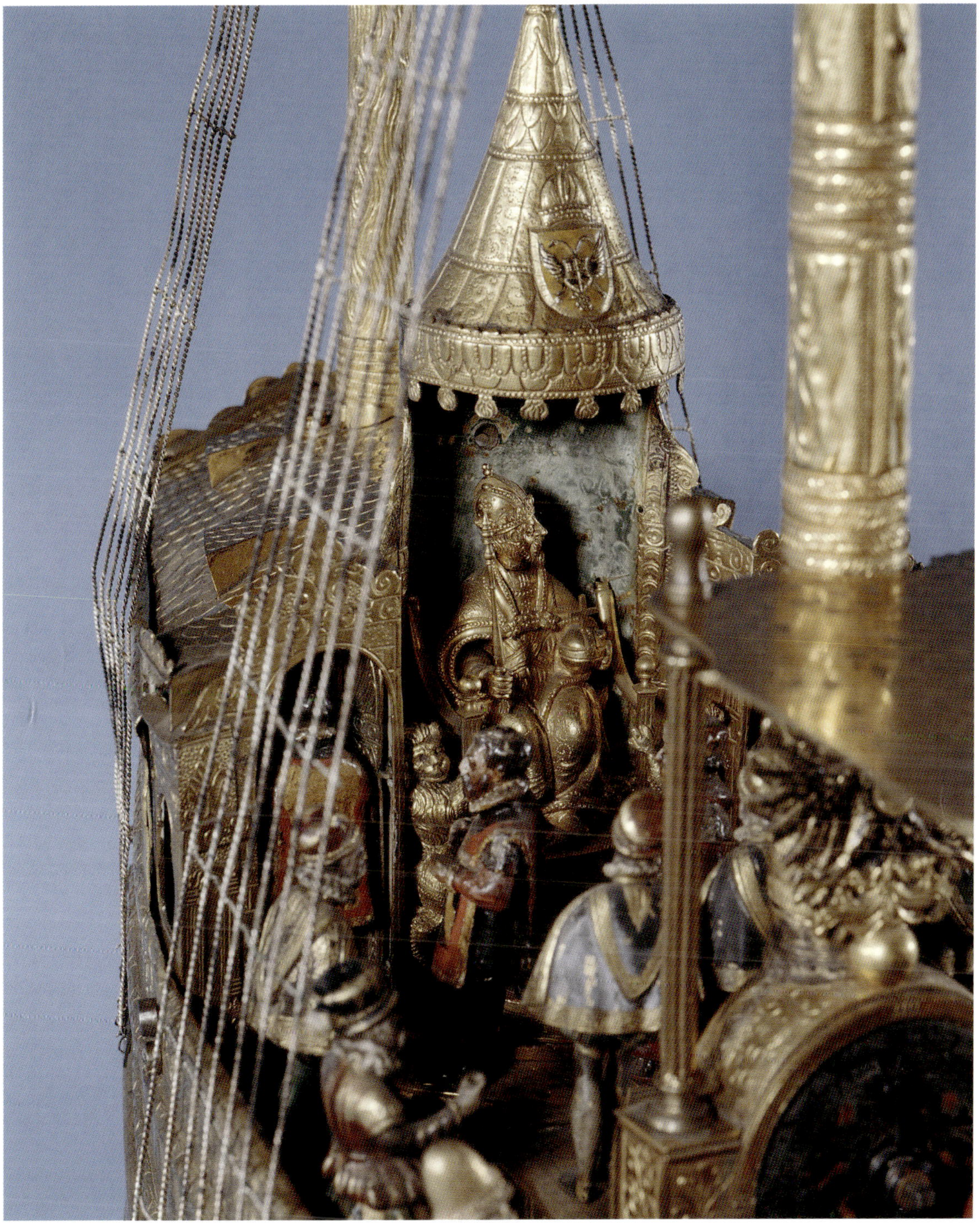

7. Hans Schlottheim, Écouen Nef (detail), 1580–1590, Musée national de la Renaissance, Écouen.

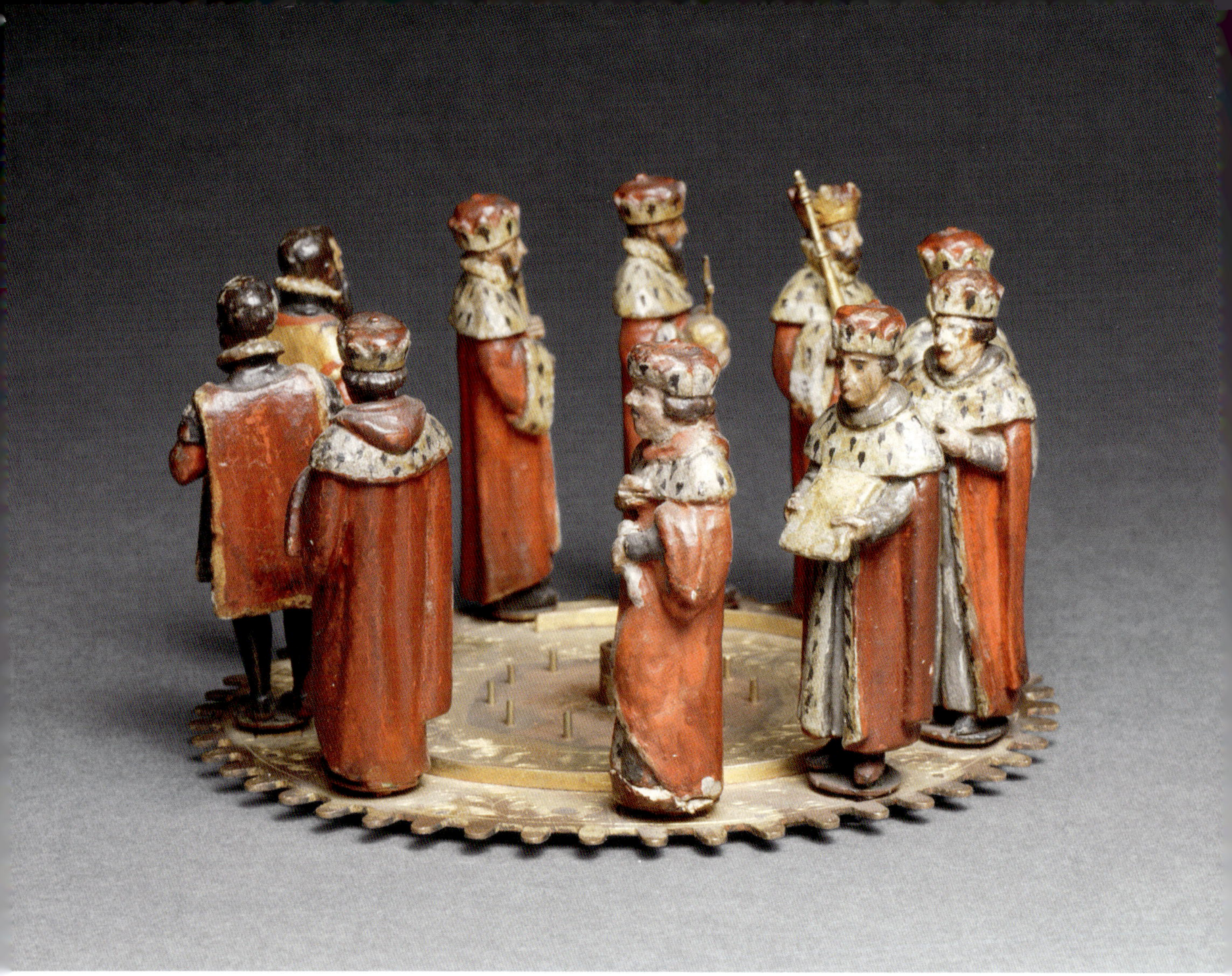

8. Hans Schlottheim, Écouen Nef (detail), 1580–1590, Musée national de la Renaissance, Écouen.

Bohemia a grand chalice, and the elector of Palatine a large key over his right shoulder.

Both nefs adhere to traditional early modern *Majestätbilder*—images in which the emperor is depicted enthroned and surrounded by his electors. Consider, for example, the woodcut *Charles V's 1520 Coronation in Aachen* (fig. 9) by Hans Weiditz (1500–1536).[34] At first glance the woodcut appears strikingly similar to the coronation scenes on the nefs; the enthroned emperor is surrounded by seven standing electors, holding either texts or liturgical objects used in the coronation ceremony. Yet Schlottheim has manipulated the Majestätbilder tradition. In Weiditz's woodcut the emperor is seated on a modest, high-backed throne draped with a simple, unornamented cloth of honor and topped with the double-headed imperial eagle. On the London and Écouen Nefs, Charles V sits on a low stool flanked by two lions.[35] The thrones on Schlottheim's works recall King Solomon's throne, which, according to the first book of Kings, was adorned with "stays on either side by the place of the seat, and two lions standing beside the stays" (I Kings 10:18–20).[36] We might conclude that the throne simply signals that the Holy Roman Empire and the emperor are extensions of the Old Testament kings and their divinely protected kingdoms—that the Solomonic throne is mobilized to convey the divinity, longevity, and legitimacy of the Habsburg monarchy.[37] Yet the inclusion of Solomon's throne here has a more potent and consequential message. For unlike Emperor Justinian (482–565), who at the dedication of Hagia Sophia allegedly cried out, "Solomon, I have outdone you!," the emperor on the nefs has not usurped or replaced Solomon but become the Old Testament priest-king. To put it another way, the emperor's pretensions to the regency of Solomon, Solomonic justice, and Solomonic wisdom receive, atop these armed galleons, a sixteenth-century formulation. On the nefs, Charles is the embodiment of kingly prudence, the incarnation of justice, the living law, the *lex animate*.[38]

"Sovereignty," writes Leszek Kołakowski, "means not being bound by external law."[39] This notion is embedded in Proverbs 8:14–16, a passage oft cited by jurists who consciously muddied the distinction between the heavenly and earthly sovereigns in order to circumscribe the power of the emperor.[40] "From me come advice and ability; understanding and power are mine. Through me kings hold sway and governors enact just laws. Through me princes wield authority, from me all rulers on earth derive their rank." It is this fierce demarcation of kings in the Holy Roman Empire that we see animated on the nefs, as the electors—the emperor's sub-kings—uniformly orbit, by way of a clockwork mechanism, the Holy Roman Emperor, the *Dominus Mundi*. It is the assigned, routine, and "ordered functioning" of the imperial governmental apparatus made visible. In this way the nefs literally, figuratively, and ideally seem to prefigure, by almost an en-

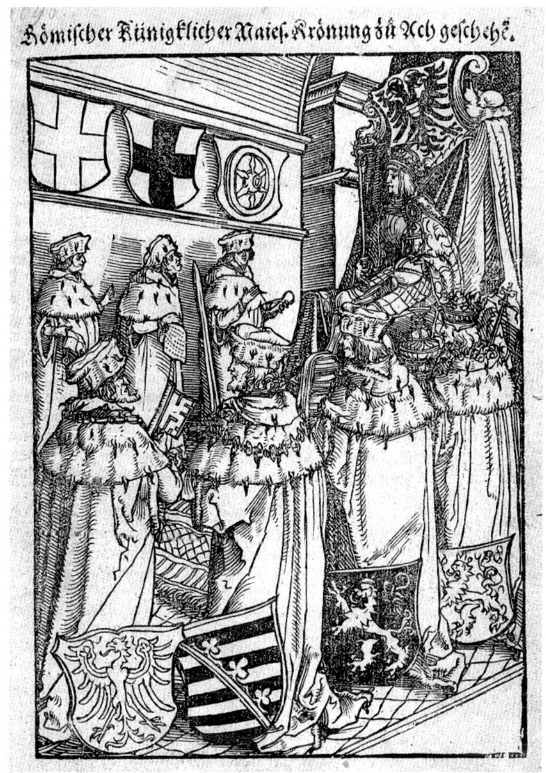

9. Hans Weiditz, *Holzschnitt zu einer vierseitigen Schrift über Einzug und Krönung Karls V. in Aachen 1520 mit Darstellung Karls V. und den drei geistlichen und vier weltlichen Kurfürsten mit Symbolen ihrer Krönungsämters*, 1500–1536, Aachen Stadtarchiv.

tire century, Nicolas Malebranche's *Treatise on Nature and Grace* (1680), in which he put forth a theory of divine government that is controlled by general laws: "Thus God executes his plans by general laws. . . . Certainly it requires a greater breadth of mind to create a watch which, according to the laws of mechanism, goes by itself and regularly—whether one carries it oneself, whether one holds it suspended, whether one shakes it as one pleases—than to make one which cannot run correctly if he who has made it does not change something in it at every moment according to the situation it is placed in."[41] Because there is an emperor, the empire runs in a constant and uniform way. The emperor reigns—he does not govern.

Yet unlike the Old Testament's legislator Solomon, who was considered a harbinger of Christ, the ruler of the Holy Roman Empire was construed as a "shadow" of Christ, an impersonator of Christ—*Christomimetes*. In the words of Ernst Kantorowicz, on earth the emperor "presented the living image of the two-natured God. . . . The divine prototype and his visible vicar were taken to display great similarity, as they were supposed to reflect each other."[42] This conclusion, *de similibus ad similia*, is made apparent when we compare the figure of Charles V on the London and Écouen Nefs to a detail of Christ as *Weltenreicher* (ruler of the world) from the *Christusmantel* (figs. 10 and 11), an embroidered liturgical pluviale and one of the key liturgical raiments of the Habsburg regalia.[43] The *Christusmantel* and the figures of Charles V on the nefs share strategies of exhibiting an imperial presence. Austere and immobile majesty is a common feature. Christ and the figures of Charles V sit frontally, iconically, in canopied thrones before concave backdrops sumptuously ornamented with vegetal motifs. The figures of Charles V and Christ are draped in dalmatics and wear mitre crowns topped with a cross. Consequently, each representation calls attention to the role of the enthroned and crowned figure as the supreme ruler of the Christian world. "Christianity," Michel de Certeau reminds us, "was founded on the loss of a body . . . the body of Jesus Christ."[44] The London and Écouen Nefs claim that the Holy Roman Emperor was a sole ruler who stepped into that void and took the helm.

Navis Ecclesiae

"In the beginning, the spirit of God moved upon the face of the waters" (Genesis 1:1–2). This pronouncement, in the opening verses of Genesis, bodied forth a profusion of nautical imagery in the Old and New Testaments. Noah, who saved terrestrial life from oblivion, drifted for forty days upon the surface of the water. Moses walked on the floor of the Red Sea, having split it to emancipate the Hebrews. The Lord heard the pleas of Jonah calling out from the bowels of the great fish. Christ glided across the face of the Sea of Galilee, and taught upon the water. Peter crossed water to be by Christ's side. "Peter called to him: 'Lord, if it is you, tell me to come to you over the water'" (Matthew 4:28). What is clear from these examples, and others that could be pulled from Corinthians, Jude, and Ezekiel, is that water functions biblically as a salvific substance and salvific metaphor. "And Jesus, when he was baptized, went straightway out of the water: and lo the heavens were opened to him, and he saw the spirit of God descending like a dove, and lighting upon him" (Matthew 3:16).

Like water, metaphors employing it are mutable, and the early church recognized this. In the place of Noah's Ark, which miraculously navigated the vanquished world, the early church inserted itself. "Even though the sea, with its waves and furies continually surges against it, God will preserve it from being submerged," wrote Augustine of the church.[45] The denotation of the central processional tunnel of a church as the "nave," *Hauptschiff* or *Kirchenschiff* in German, bears witness to the power and longevity of the metaphor. Early church fathers such as Augustine

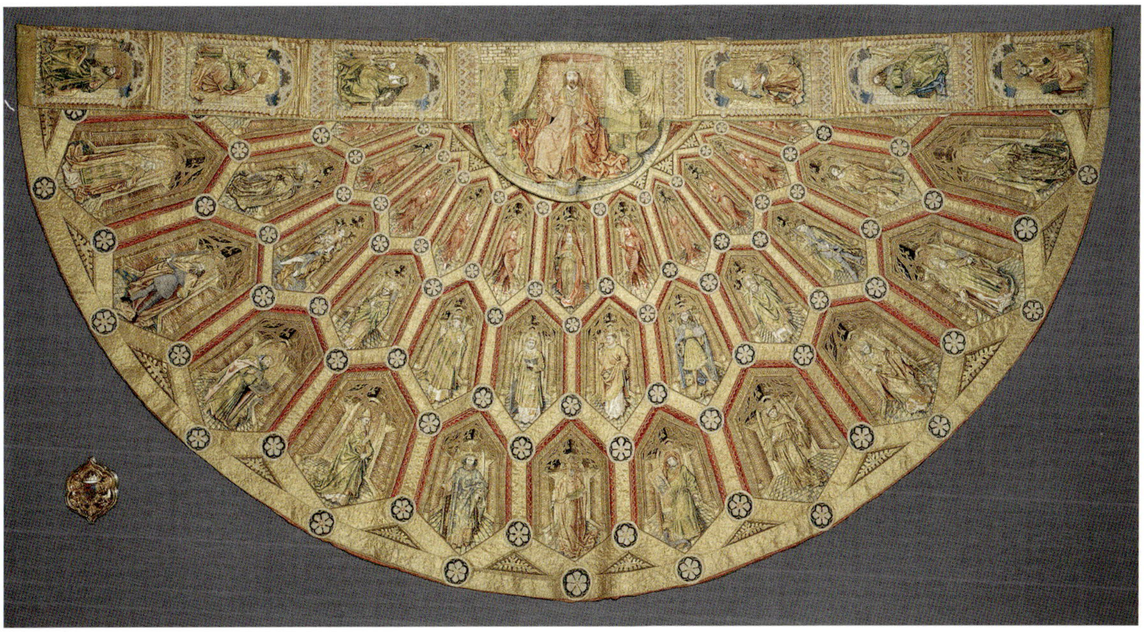

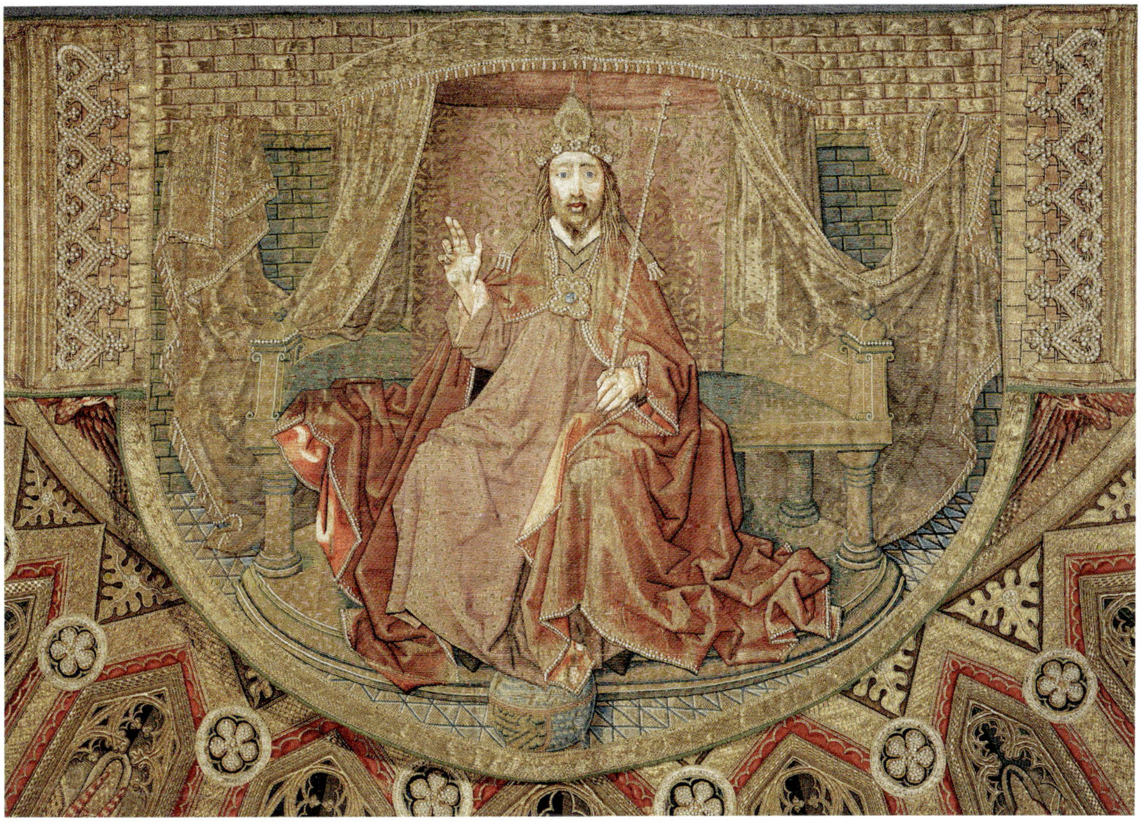

10. *Christusmantel*, 1525, Weltliche Schatzkammer, Vienna.

11. *Christusmantel* (detail), 1525, Weltliche Schatzkammer, Vienna.

and Ambrose, and later theologians such as Remigius von Auxerre, Pseudo-Clement, and Richard of St. Victor, made clear that the captain of this vessel fortified and endowed with divine favor was the pope.[46]

A late fifteenth-century Italian engraving of an encounter between Pope Paul II and Holy Roman Emperor Frederick II, which took place between December 24, 1468, and January 9, 1469, shows that the identity of the captain of the Navis Ecclesiae was far from evident (fig. 12). Presenting a medley of heraldic and allegorical images and texts that signify or allude to European nobility and principalities, the engraving meticulously diagrams the political relationship between the emperor and the pope.[47] That the heavens predestined the meeting is made apparent by the burst of light from the comet of 1468 and a text, running along the base of the image—"The King of the Romans will Claim the whole Christian Empire for himself"—that refers to the prophecy of the Tiburtine Sibyl.[48]

According to the Tiburtine Sibyl, Christ's second revelation would occur upon the defeat of the heretic by the last Christian Emperor.[49] Eschatological in scope and promissory in nature, the prophecy helped to solidify the Roman Empire into a temporal monolith. It was Eusebius (260/65–339/40), court historian of Constantine the Great (227–337), who saw the shape of human history as bounded by the birth of the Roman Empire. In his *Commentary on the Psalms* he wrote: "When the Lord and Savior appeared and, at the same time as his advent Augustus, first of the Romans, became King of nations, pluralistic polyarchy was dispersed and peace covered all the earth."[50] For Eusebius, as well as John Chrysostom, Prudentius, Ambrose, Jerome, and even the later Orosius, there was a correspondence between the unified global empire and Christ's birth, a correspondence between monarchy and the first revelation of God.[51] The Tiburtine Sibyl erected the second boundary. For the Tiburtine Sibyl the correspondence lay in the last Christian emperor's reunification of the empire (by way of obliterating all nonbelievers) and Christ's second revelation—the advent of God's kingdom. The biblical corroboration that Christ's second coming would not immediately follow the first is 2 Thessalonians 2: "Now we beseech you brethren, by the coming of our Lord Jesus Christ, and by our gathering together unto him, That ye be not shaken in mind, or be troubled, neither by spirit or by word, nor by letter as from us as that the day of Christ is at hand. Let no man deceive you by any means: for that day shall not come, except there come a falling away first and that man of sin be revealed, the son of perdition." Only the second coming's suspension enables the Christian empire to come into being. The empire is at once existentially dependent on the deferral of this apocalyptic event and the agent that brings it about.

But the Tiburtine prophecy is just a prompt in this engraving, because through its prolixity, the image divulges the structure and indignity of imperial and papal machinations. Who, asks the image, is the head of the Christian empire and the captain of the Navis Eccleasiae? The answer is far from clear-cut, as we see from the physical struggle between Frederick II and Paul II, who are stripped of their princely garb. Each potentate has a foot planted on the mast of the ship, while his other foot rests on a less reliable support: for Frederick it is the cringing lion of Burgundy; for Paul, the wheel of the patriciate of Rome. The monarchy of France, here embodied by the heraldic shield of three fleur-de-lys, has been, twice even, pushed to the margins. Enjoying a special status are the dukes of Austria (the Habsburgs), not merely because "DVCES AVSTRIE" is centrally placed on the vessel but because the line attached to the anchor, which keeps

12. *Allegory of the Emperor and the Pope*, ca. 1495, National Gallery of Art, Washington, D.C.

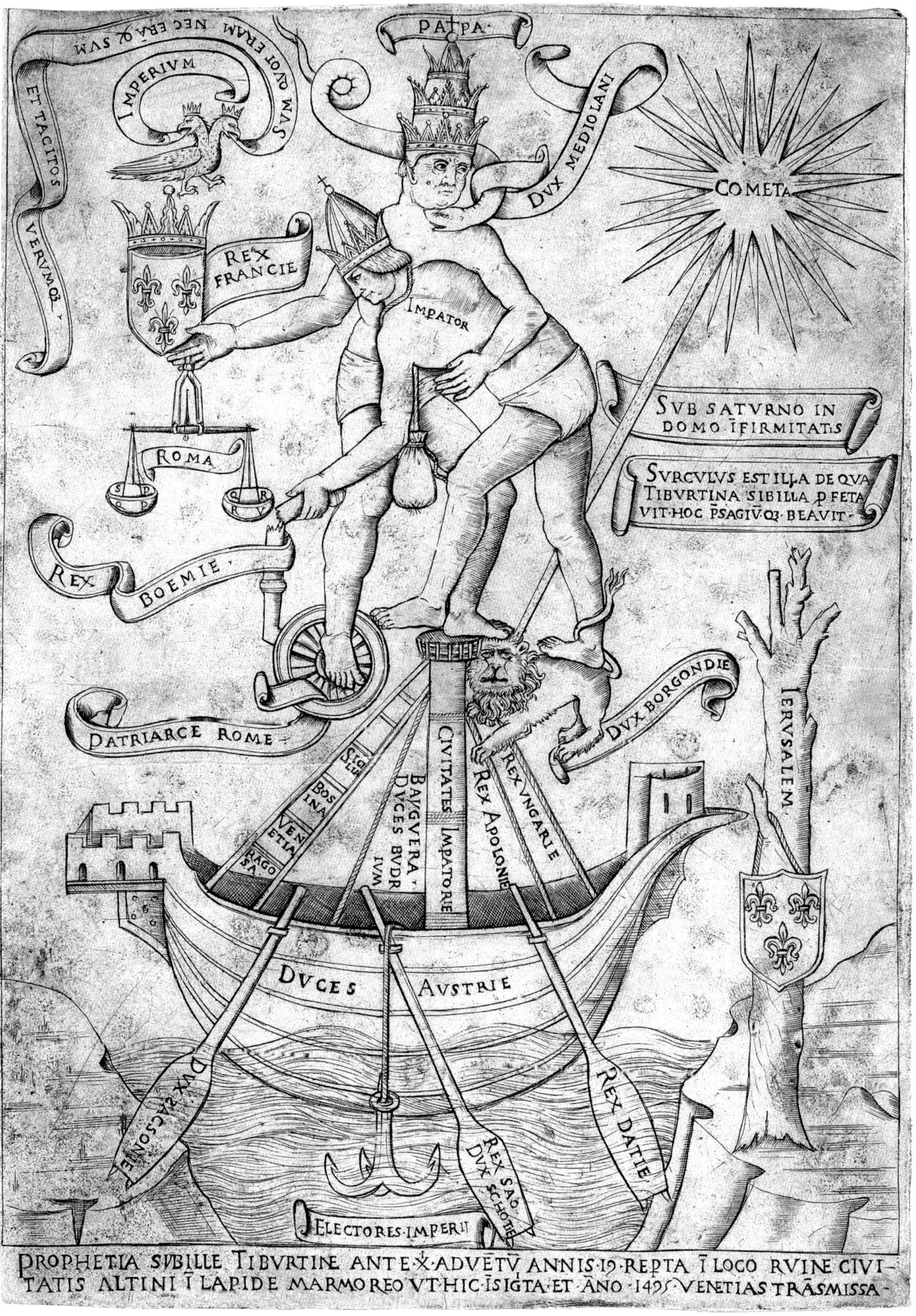

PROPHETIA SIBILLE TIBVRTINE ANTE X ADVETV ANNIS 19 REPTA I LOCO RVINE CIVITATIS ALTINI I LAPIDE MARMOREO VT HIC ISIGTA ET ANO 1495 VENETIAS TRASMISSA

the boat-church from drifting, bisects the two words.

Whereas the engraving is pessimistic and ambiguous about the identity of the leader of the Christian empire and the Navis Ecclesiae, the London and Écouen Nefs are optimistic and explicit. It is Charles V, and eventually the heir to his throne, Holy Roman Emperor Rudolf II—the patron of these objects. Why, however, would a representation of Charles V's coronation have been desirable for Rudolf II to begin with? Why would the emperor not wish to memorialize his father, Maximilian II (1527–1576), or perhaps the first Habsburg ruler—Rudolf II's namesake—Rudolf I (1218–1291)? A partial answer is to be found in Charles V's reign and dynastic intentions. A more complete answer lies in the shaky dynastic grounds on which Rudolf II ascended to the imperial throne, since he was not, technically, Charles V's heir. Rudolf's cousin, Philip II of Spain, (1527–1598) was, and Rudolf II took the crown in spite of Philip II's prolonged and virulent protests.[52]

Viewed against the backdrop of the Habsburg succession issue, the London and Écouen Nefs can be seen to have legitimized the Austrian line of the house of Habsburgs generally and Rudolf II in particular by representing the coronation of Charles V. Through these objects Rudolf could visually tie his imperial title to Charles V, as if he had directly inherited the throne from him. This linking of Rudolf II's dynastic heritage to Charles V was an active political strategy, reiterated visually and pictorially in several instances. On May 7, 1577, Rudolf II ordered that Charles V's coat of arms be displayed in the Schlosskirche (the palace chapel) and the church of Saint Jacob in Prague.[53] Four years later, the emperor sent Philip II 117 gulden and 20 kreuzer for a portrait of Charles V that was in Spain.[54] And finally, the emperor juxtaposed the bronze bust of himself, now known as *Portrait Bust of Rudolf II* (ca. 1590, Vienna, Kunsthistorisches Museum), by Adriaen de Vries, with Leone Leoni's *Bronze Bust of Charles V* (ca. 1540, Vienna, Kunsthistorisches Museum) in the imperial Kunstkammer in Prague.[55]

The Eternal Recurrence

On the day of his anointment Charles was paraded through the streets of Aachen. When the emperor-to-be reached the city's cathedral, built by Charlemagne himself, he was met by the archbishops of Mainz and Trier outside the entrance. The two Geistliche Kurfürsten flanked Charles as the door opened, and the music of the imperial guard came to a halt while the sacred music of the liturgy poured out of portals of the church. Charles was escorted by the two Kurfürsten to the high altar, where he was asked if "he would swear to preserve the ancient faith, protect the church, govern justly, and care for the humble, poor, widows and orphans."[56] Charles answered "Volo" (I will) to each question posed to him. After Charles's affirmative answers, the archbishop of Cologne turned to the congregation and asked if they would pledge their undying allegiance to Charles as their emperor. Charles was then anointed on the back of his neck, his chest, his hands, and his head while the archbishop of Mainz repeated, "Ungo te regem oleo sanctificatio." Once Charles had been mystically distinguished from those on Earth and placed under the authority of God, he was crowned with the *Coronum Caroli* (crown of Charlemagne) and given the scepter and globe while the electors circled him as the congregation chanted, "Vivat, vivat, vivat rex in aeternum."[57] This political and liturgical rite of coronation legitimated Charles V, demonstrated the allegiance of the empire and electors and the continuity of Charlemagne's empire with that of the present, and fortified the Habsburg line.[58]

Given the questions about Habsburg succession that troubled Rudolf II, it makes good sense that the symbolic framework of this momentous occasion should have been taken up by Schlottheim in the London and Écouen Nefs while he was working at Rudolf II's court.[59] When viewing the nefs, we behold not simply a documentation of Charles V's coronation, as we did in Hans Weiditz's woodcut, discussed earlier, but rather a re-creation and re-performance of three imperceptible metamorphoses to which the ritual gives material form: the transformation of a man into a sovereign; the transformation of a man into the living law; and his transformation into a placeholder for Christ. Furthermore, whereas we might conclude that the creation of a novel ritual site atop a galleon might have undermined the ideological and sacred significance of the event, it in fact did the opposite. This is no ordinary ship, for it evokes "the ship God has built to navigate in the world's flood." With the help of the object's internal mechanism, the coronation—an ephemeral act in the extreme, and a precarious one to execute—becomes more reliable, durable, and repeatable. It is as if Rudolf chose to convey that the process of succession and the maintenance of the Holy Roman Empire ran like clockwork, by way of a clockwork object.

More Variations on the Imperial Theme

In his continued pioneering and indispensable work on the Rudolfine court and Rudolfine art, Thomas DaCosta Kaufmann has established that the imperial conceit of Rudolf II—a conceit that suffused imperial collecting practices, imperial art production, and imperial ceremony and festivities—was one of world domination.[60] Kaufmann has shown that this was so not only because a number of humanists and poets fashioned Rudolf to be the lord of the universe, but also because such an idea was on the face of the works of art created under Rudolf's patronage. Perhaps no better example of this painstakingly crafted imperial image is Giuseppe Arcimboldo's portrait of Rudolf II as Vertumnus, the Roman god of the seasons (fig. 13). Painted in Milan around 1590 and likely sent to the imperial court along with several poems by Gregorio Comanini and Giovanni Filippi de' Gherardini in January 1591, this *portrait historié* tests the capacity of the imperial image to be represented directly and indirectly.[61] For here the sitter is both himself *ex se*, out of himself, as well as something beyond himself.[62] In *Vertumnus*, Arcimboldo builds an image of a crowned Rudolf II out of an array of flowers, fruits, grains, and vegetables representing all the seasons. "This undifferentiated conjunction of seasons," writes Kaufmann, "suggests that they form one undifferentiated season: this is the mark of the Golden Age that has come with the reign of Rudolf. This Holy Roman Emperor is thus shown as the ruler of a new Rome that will enjoy the eternal domination of the world."[63]

Here, portraying the ruler of the new Rome is, as Roland Barthes noted, a poetic activity. Of Arcimboldo's imagination Barthes wrote: "It does not create signs, it combines them—deflects them—exactly what a craftsman of language does."[64] Indeed, Vertumnus is an extraordinary example of an artist translating a variety of objects into a convincing human likeness. Bulbous fruits and squashes—a cucumber, a pumpkin, an artichoke, a head of garlic, a turnip—dominate the composition, creating the undulating surface of the figure who emerges from the black ground. Heavily modeled and outlined, each piece of fruit or vegetable becomes a unique object carefully and systematically fitted together. The apples that stand in for Rudolf's cheeks, the wheat for his beard, the peas for his eyebrows, and the berries for his pupils make Arcimboldo's likeness of Rudolf celebratory, because it articulates a very particular oscillatory relationship between the figure represented and nature: an oscillation between the macrocosm and the microcosm that authorizes the power of the emperor. Moreover, all these fruits of the earth, read together, describe an eternal cycle of renewal. They are, as Kaufmann has elucidated, an intrinsic part of the empire's image of itself, Arcimboldo representing the bounty of the world as the emperor and the emperor as the

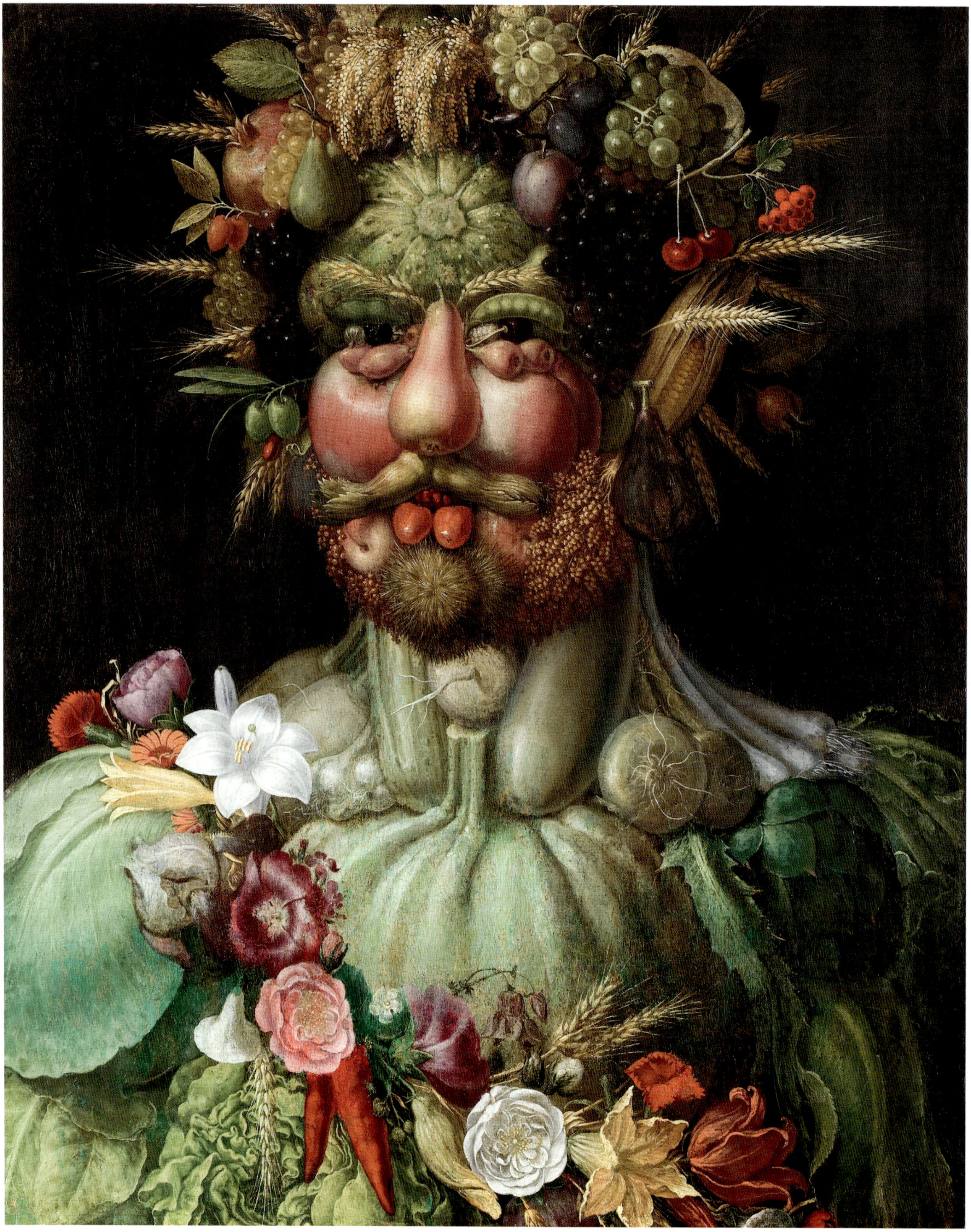

ultimate bounty of the world. Portraiture becomes allegory and allegory becomes portraiture. The point of the portrait is that the emperor is isolated from any source of political authority, divine, historical, or human.

Whereas the *Vertumnus* painting creates a fiction of imperial autochthony, the London and Écouen Nefs argue that imperial authority is extrinsic. Instead of conveying an imperial ideology of a Golden Age through allusions to mythical and literary figures that drew upon late sixteenth-century systems of metaphorical and metaphysical correspondences "according to which . . . the macrocosm of the universe parallels the microcosm of man,"[65] the nefs refer back to and laud an earlier notion of the emperor and imperial sovereignty through biblical, ecclesiastical, and liturgical imagery, and by doing so make clear that the emperor's power, his right to rule, is bestowed upon him by sacred authorities external to him.

Both nefs not only performed and *re-performed* the sacramental and liturgical event that transformed Charles V into a *persona gemina*; they also emphasized the emperor as the earthly representative of God through his association with Old Testament kings and Christ. To put it slightly differently, as representational devices the nefs provided a place wherein the act of consecration and the conviction of the emperor's role as Christ's vicar on earth could be celebrated and reified. The differences between these two modes of representing and energizing the emperor and empire—one allegorical, the other liturgical—reflect the range of imperial themes at the court of Rudolf II. Each object, or each mode, is a unique deposit of a political statement. These unique deposits, however, are not necessarily inimical to each other but instead suggest that the transition from the medieval to an early modern notion of empire and kingship was difficult, and that this issue was a live one at the court of Holy Roman Emperor Rudolf II.

13. Giuseppe Arcimboldo, *Vertumnus*, 1590, Skoklosters Slott, Stockholm.

But let us return now to the more concrete political stakes of the London and Écouen Nefs. As I suggested earlier, in addition to stressing the apostolic legitimacy of Charles V's reign, the automata benefited Rudolf II by implication. By commissioning such objects, Rudolf II directly associated himself with the coronation of Charles V, as if no problematic dynastic issues haunted that association. Consequently, both automata may be viewed as attempts to legitimate Rudolf II's tendentious claim to being one of the legitimate progeny of Charles V. By foregrounding and emphasizing Charles V's coronation, the automata effectively push the contested coronation of Rudolf II into the background and allude to the licit stronghold of the Austrian line of the House of Habsburg over the Spanish line, whose head was Philip II—Charles V's only son and heir.

The imperial problems that the nefs addressed were real and had implications beyond the court in Prague. It is my considered opinion that the automaton listed in the 1587 inventory of the Dresden Kunstkammer with which this chapter opened is in fact the London Nef, and that Rudolf gave the London Nef to the elector of Saxony, Christian I (1560–1591). Although we have no archival documentation of this gift, it would have been a rather appropriate present for Christian, a Protestant, when he took the title of elector in 1586, as the nef clearly represents the ideal relationship between the emperor and his electors. Rudolf's treatment of Protestants and Protestant princes in the empire was mixed. In 1578, two years after he took the throne, he actively rooted out Protestants in Vienna. Three years later, in 1581, he banned Protestant worship in Aachen, the site of Charles V's coronation and, historically, the site of ecclesiastical power in the Holy Roman Empire. When he convened his first imperial Diet in Augsburg in 1582, the same Diet in which he commissioned the two nefs, the Protestant electors—of Saxony, Brandenburg, and Palatine—did nothing to stop his attempts to align his faith with that of his subjects.[66] Thus we might view the nef as a reminder to the Protestant elector of his position vis-à-vis the holder of the imperial throne.

There are, however, other plausible explanations for this particular offering. Perhaps Rudolf was self-conscious about the new elector's opinion regarding the legitimacy of his imperial crown. Or perhaps there is a much larger issue staring back at us. The court in Dresden was a Protestant one and, as such, could not theologically accept that the Catholic emperor could be installed as Christ's vicar on earth. Certainly the Protestant elector could not accept that the ceremony of the coronation was anything but, to use Erasmus's own words, "sublime stupidity." If Rudolf did give the London Nef to Christian I, it was but the first automaton with a potent message he received as a gift. We now turn to the second.

Three

The Gifts That Keep on Giving

On New Year's Day in 1589 Sophie of Brandenburg (1568–1622), the wife of the newly instated Albertine elector of Saxony, Christian I (1560–1591), gave her husband a custom-made automaton featuring an enactment of the adoration of the three kings (fig. 14).[1] Crafted by the Augsburg goldsmith Hans Schlottheim, the maker of the nefs discussed in the previous chapter, the automaton was an elaborately decorated, multitiered construction of bronze that was gilded and embellished with silver; by way of an internal mechanism it played two consecutive hymns while the shepherds and kings of the Christmas story moved in a circle presenting gifts to the Christ child. The automaton was destroyed during the Allied bombing of Dresden on February 13, 1945. Although now lost, the automaton has attracted some scholarly attention and is commonly referred to as the Christmas Crib Automaton.[2]

The Christmas Crib Automaton was singular in numerous respects. First, it appears to have been the only automaton crafted by Hans Schlottheim representing the adoration narrative, whereas Schlottheim tended to make automata en masse. His famed nefs in the British Museum and the Musée national de la Renaissance in Écouen exemplify his tendency to reuse models and technology over the course of a decade (figs. 2 and 3). Second, the Christmas Crib Automaton was commissioned by a wife as a gift for her husband. In the second half of the sixteenth century, automata were generally given as tokens of allegiance between male rulers; they were diplomatic gifts. In the mid-1560s the bishop of Augsburg, Otto von Truchsess, presented an automaton to Holy Roman Emperor Maximilian II. This vessel, discussed in chapter 2, was made completely of silver and depicted an emperor enthroned in the presence of the seven electors of the Holy Roman Empire. And recall that in 1582 the duke of Bavaria, Wilhelm V, gave Archduke Ferdinand II of Tyrol an automaton with Bavarian trumpeters playing a processional song, while the ebony base on which they stood displayed the mingling of the Bavarian and Austrian coats of arms, visually and audibly marking a celebratory connection between the Wittelsbach and Habsburg Houses (fig. 1).[3] Sophie's gift of an automaton to her husband was not in keeping with this established tradition. There is a third peculiar aspect of the Christmas Crib Automaton. It was self-reflexive. In the manner of other early modern gifts, it "present[ed] to the viewer both the present and the supposed spirit in which it was given."[4]

The Christmas Crib Automaton was given by Sophie to her husband during the New Year's festival, which in the late medieval and early modern eras abounded with banquets, games, and displays of largess. The period between Christmas and New Year's was filled, at both Protestant and Catholic courts, with visitations of foreign dignitaries, the burning of logs, the performance

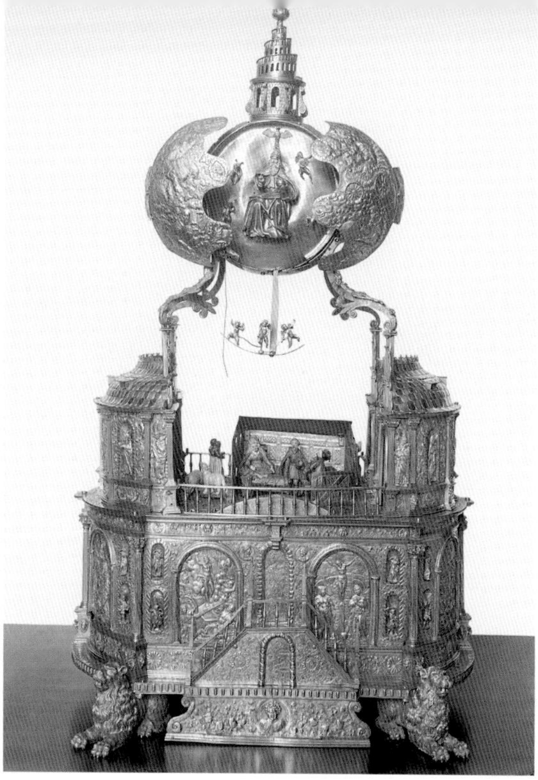

of Christmas plays (*Weihnachtsspiele*), the consumption of food, and the giving of gifts.[5] The Christmas Crib Automaton was one of two objects in an exchange of gifts between Sophie of Brandenburg and Christian I. Christian reciprocated Sophie's gift with a *Prunkkassette* crafted by the famed and recently deceased Nuremberg goldsmith Wenzel Jamnitzer (fig. 15).[6] Initially the gifts appear to adhere to conventions of courtly prestation. Christian's gift is consistent with a long history of men presenting to women cases to hold their jewelry, barrettes, combs, mirrors, and maquillage—objects that enabled women to adorn and beautify their bodies.[7] And Sophie's gift of an automaton was seemingly well suited for a prince. The remarkable craftsmanship by Schlottheim and the wondrous movement of the object was, on one level, intended to delight Christian. And yet on another level, as this chap-

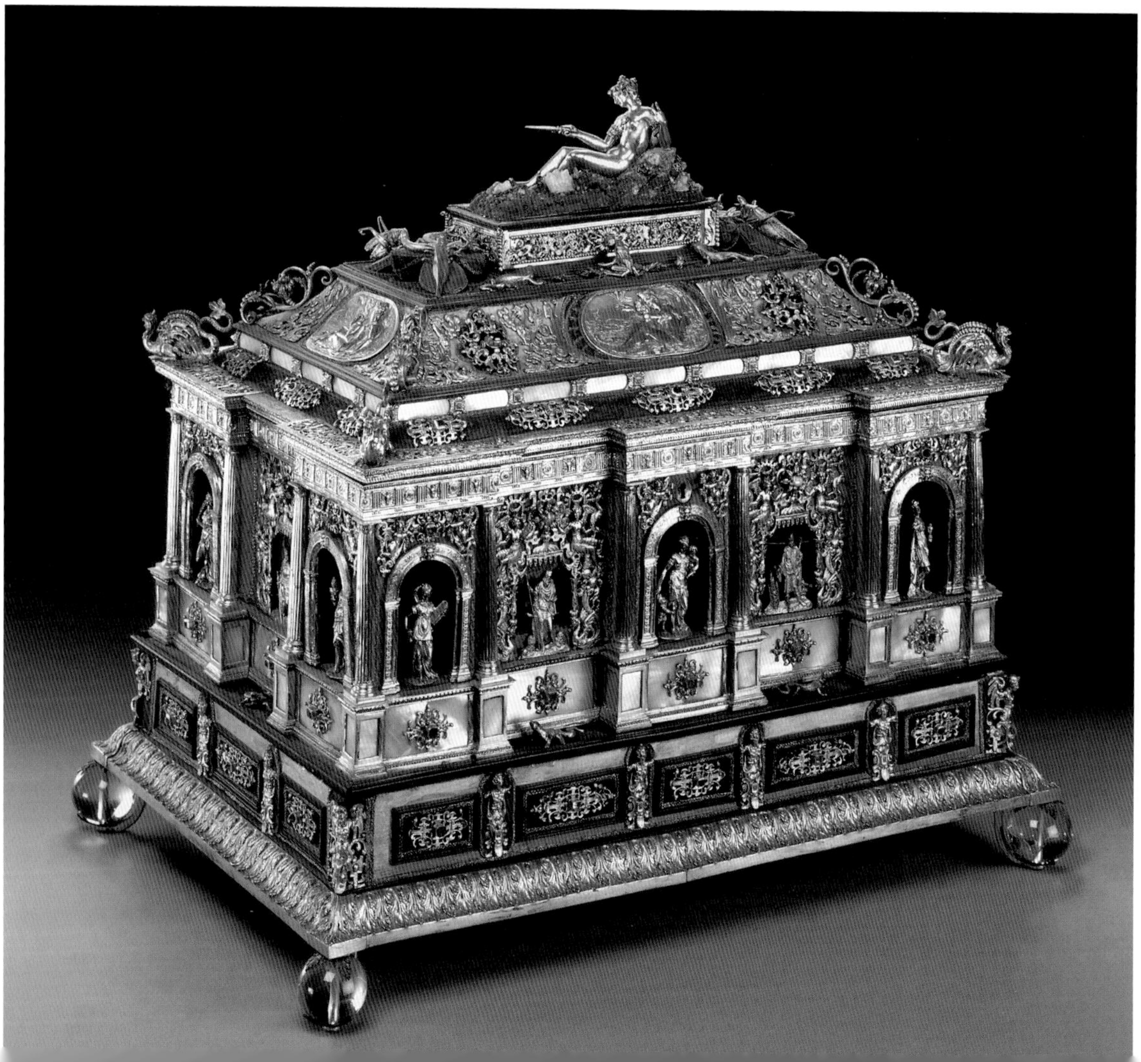

ter argues, Christian's mirth was not the only aim.

The Christmas Crib Automaton was a gift whose central action was giving and whose central theme was recognition—the recognition of Christ and, as we shall see, the self-recognition of its intended recipient. It presented Christian I with a model of proper actions and attitudes of a Lutheran prince. In this way it was not unlike a *Fürstenspiegel* (mirror of princes), a text intended to instruct a ruler on good governance and appropriate behavior, such as Erasmus of Rotterdam's *Institutio principis Christiani* (1516). But the Christmas Crib Automaton was a model in two other, more specific senses. It provided a concrete and simple example of a complex idea, and it anticipated the future by encouraging something to be derived from it. In other words, rather than affirming the status quo, it proposed an alternative.[8] This interpretation of the Christmas Crib Automaton and its political and religious significance is also largely based on an elaborate mirror that Sophie of Brandenburg gave to her son Elector Christian II (1583–1611) in 1600, the Lüneburger Spiegel of ca. 1587–92 (fig. 16). This gift, a masterpiece of metalsmithery, placed the ruler of Saxony at the center of an overtly political allegory. Like the Christmas Crib Automaton, the Lüneburger Spiegel was steeped in a Lutheran tradition grounded in Martin Luther's writing, was given by Sophie of Brandenburg at a critical political moment, and displayed its intended recipient as a virtuous Lutheran ruler. The mirror accomplished this through a similar visual dynamic but to a very different end.

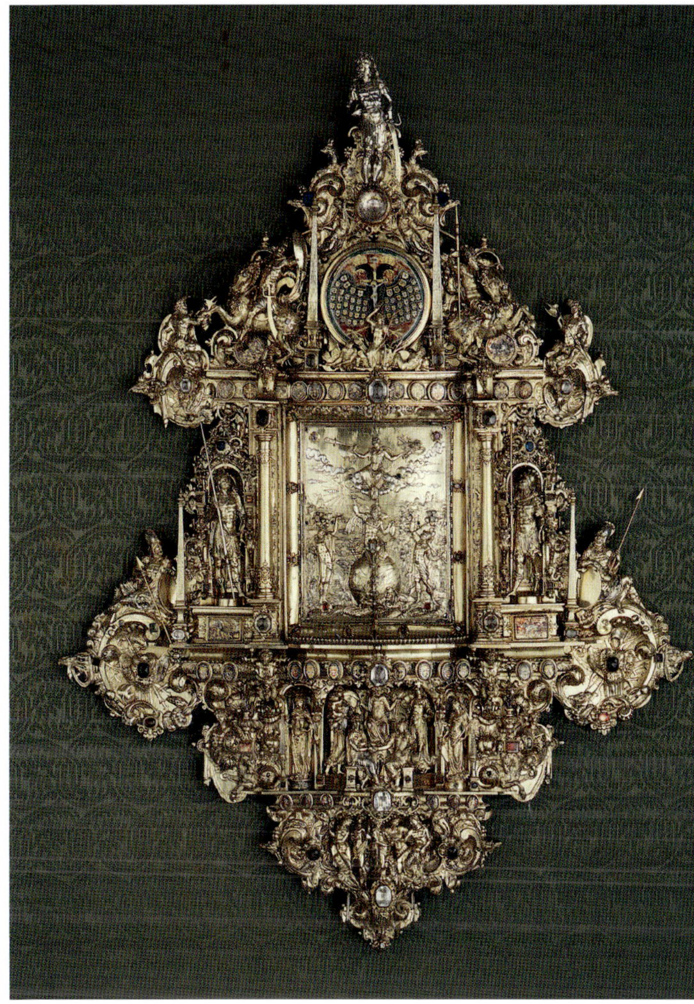

16. Luleff Meyer and Dirich Utermarke, Lüneburger Spiegel, 1587–1592, Grünes Gewölbe, Dresden.

14. Hans Schlottheim, Christmas Crib Automaton, 1588, formerly in the Mathematisch-Physikalischer Salon, Dresden.

15. Wenzel Jamnitzer, Prunkkassette, 1588, Grünes Gewölbe, Dresden.

Protestant Rulers in Saxony

Saxony was the first Protestant territory in the Holy Roman Empire, as well as the most influential. It was throughout and after the reign of Christian I's father, Elector August I (1526–1586), that strong Protestant marriage politics began to play a larger role in Saxony's maintenance of Protestant power.[9] In 1548 August married Anna, princess of Denmark (1532–1585), daughter of Saxony's strong Lutheran ally Christian III (1503–1559). The nuptial contract both expressed and reinforced the Kingdom of Denmark's and Saxony's mutual commitment to Lutheranism.[10] This is not to deny that August would manipulate marriages to uphold amicable relationships with Calvinists outside Saxony. August made a strategic move in 1570 by marrying and exporting his daughter Elisabeth to Johan Casimir, the elector of Palatine.[11] At home, though, August took great measures to assure that his dynasty in Dresden presented itself as the bulwark of Lutheranism. August married his son and heir Christian I to Margravine Sophie of Brandenburg in 1582, thus grafting the Albertine house to the small, but up-and-coming, staunchly Lutheran court in Cölln (now Berlin).[12]

Johann Georg I (1525–1598), the elector of Brandenburg and Sophie's father, had a zeal for orthodox Lutheranism. In 1572, ten years before Sophie and Christian married, Johann Georg issued a new church ordinance that was intended to strengthen Brandenburg's Lutheran church and cleanse it of its ceremonial traditionalism. The ordinance stated that God's word alone was to be preached, that Luther's own writing was to be read diligently by his subjects, and that his subjects were to avoid all other books and teachings. It follows that after 1572 Johann Georg refused to acknowledge any other reformed ideas in his lands.[13] The year 1573 saw the publication of the Visitation and Consistory Ordinance, which proclaimed that all ministerial candidates must demonstrate their command of the Augsburg Confession of 1555 and Luther's writings, in addition to Scripture, before they could be ordained and recognized by the state. By institutionalizing the organization and composition of the Lutheran Church in Brandenburg, and by expelling any subject whose religious views and practices did not, in his view, hit the mark, Johann Georg attempted to enforce confessional uniformity in Brandenburg. Thus, it must have been quite advantageous for Johann Georg I to send off one of his thirteen daughters to eventually rule over the most stalwart Lutheran province. It not only afforded his family, the Hohenzöllerns, the opportunity to put themselves on the political map, but it also gave the House of Brandenburg a strong religious and political alliance within the empire.[14] That this partnership was prized and vaunted is attested to by medals struck to commemorate the unification of the houses of Saxony and Brandenburg. One such medal in the Münzkabinette in Dresden, a self-consciously classicizing object, portrays Johann Georg and his wife, Elisabeth of Brandenburg, seated together in profile on the obverse, while on the reverse we find in the same position Christian I and Sophie of Brandenburg. The medal demonstrates that the two couples are not just related but interchangeable, almost indistinguishable. Johann Georg is Christian and Christian is Johann Georg.[15]

The gift of the Christmas Crib Automaton must be viewed within this context of Lutheran marriage politics, in which marriage alliances helped to establish and further Protestant and specifically Lutheran agendas in Saxony and Brandenburg. Even though Sophie of Brandenburg, the giver of the automaton, was absorbed into the Albertine household, the interests of the Brandenburg court and its ardent Lutheranism were not abandoned at the altar. In fact, Sophie was famously devout. For instance, she relieved her children's tutor, Elias Reinhardt, of his duties for not providing a strict Lutheran education. Under Reinhardt, her son and future elector Christian II was responsible for memorizing all of Luther's catechism, all of the psalms, Luther's Sunday Gospel readings, and all the prayers Luther had written himself. Sophie demanded that her sons' Lutheran education exceed Reinhardt's curriculum.[16]

While the subject matter of the Christmas Crib Automaton was surely inspired by the time of year at which it was given, it should not be interpreted as a mere moving illustration of the practice of gift giving during the Christmas and New Year festival.[17] The object was commissioned and given during a time when there was a great amount of suspicion and anxiety among Lutherans, in and outside Saxony, that Christian I had secretly converted to Calvinism. And the automaton made for Christian I may have responded to these concerns.

The Albertines' complicated relationship with Calvinism went back to Christian I's father, August.[18] Even though August made valiant attempts to export and import Lutheran brides, he surrounded himself with Calvinist and Huguenot sympathizers. Hubert Languet, one of August's foreign diplomatic representatives, had fled France in fear of his life because of his Huguenot leanings. August's chaplain, Christian Schütz, had strong ties to Calvinism as well.[19] In 1574 the Lutheran clergy in Saxony attacked what they perceived as August's lack of evangelical clarity. August responded to the criticism and threw Languet and Schütz into prison, along with his advisor Dr. Georg Craco and his physician Caspar Peucer, for their Calvinist leanings.[20] August's crackdown on Calvinists at court had begun. August's betrothal of Christian I in 1582 to a woman from an extremely orthodox Lutheran court was consistent with his reactionary behavior. The marriage signified to the empire that the electors of Saxony were unwavering Lutherans. The marriage alliance with the Brandenburg court promised to alleviate suspicion about the steadfastness and orthodoxy of August's beliefs. Yet August's attempt to rectify the situation may have come too late.

Once Christian was of age, in 1584, his father slowly began giving him political responsibilities.[21] Very soon thereafter Christian required advisors, and he brought Dr. Nikolaus Krell (1551–1601) into the fold. Krell was a Calvinist, and soon started rising in the ranks. Christian made Krell his privy councillor in 1586, the year that Christian took over the position of elector, and three years later the chancellor.[22] The promotion of a Calvinist upstart elated Calvinists while again deeply offending Lutheran clergy. But Christian did not stop there. He singlehandedly paved the way for Calvinists to become faculty in Saxon universities, which were at that time living monuments to Luther; he allowed Calvinist books to be sold in Leipzig; and finally, he decreed that exorcism would no longer be a part of the baptism service—a key component of the Lutheran rite. Fears of *cuius regio, eius religio* were rampant.[23]

The Christmas Crib Automaton was commissioned by Sophie and given to Christian at the height of his very public Calvinist sympathies. It is fitting, therefore, that the automaton should have emphasized Christian's role as a Lutheran prince, especially on a visual and aural level. He would have watched the figures on the automaton repeatedly perform their action in harmony with the music, including one hymn by Luther himself. The Christmas Crib Automaton presents a repetitive image—the offering of gifts from rulers to the humble, destitute Christ—structured to reinforce the values and actions of a pious Lutheran prince.

The Act of Giving

The figures of the Christmas Crib Automaton moved in a circle atop an ornamented ovular silver base. The figures, biblical scenes, and architectural elements that populated the base were rendered in a variety of techniques that only the most adept goldsmith could have undertaken. Small figures, which were cast in the round, stood in recessed niches on each side of eight modeled arches. Each arch contained one biblical scene in relief: four from the Old Testament (the creation, the flood, Abraham's sacrifice of Isaac, and Moses before the burning bush) and four from the New Testament (the baptism of John, Christ in the Temple, the crucifixion, and the ascension). The spaces between the niches and the arches were filled with swirling vegetal motifs executed in relief and engraved detail. The base had four distinct sides. With the exception of the front of the base, each side contained two arches with biblical scenes, eight figures in niches—arranged in pairs on each side of the biblical scene—and four Corinthian columns. On the front of the object, a third, smaller arch was situated between the two larger arches, two round bosses of unidentified male heads, and only four figures in niches. Additionally, the front had a double staircase, which protruded from the base. The purpose of the staircase was to hide several mechanisms enabling the automaton's movement.[24]

The narrative flow of the biblical scenes moved from right to left, the same direction as the figures of the automaton. The first scene, the creation, appeared in the right arch on the left face of the object and was followed by the flood. The third biblical scene, Abraham's sacrifice of Isaac, was in the right arch on the rear side of the base, and Moses before the burning bush was in the left arch. Moving around the object, the next scene was the baptism of John, paired with Christ in the Temple on the right face of the base. The biblical narrative concluded on the front face with the crucifixion and the ascension, strategically placed directly beneath the figure of the Christ child in the Christmas story, thus dramatically juxtaposing the beginning of Christ's life and his voyage to paradise. It is clear that Schlottheim took the circular movement of the figures of the automaton into account when organizing the aptly chosen biblical scenes, all of which have typological significance with reference to the life of Christ.

The music that sounded from the base of the Christmas Crib Automaton consisted of well-known Lutheran hymns dedicated to the story of Christ's birth. The first was Luther's "From Heaven above to Earth I Come" ("Vom Himmel hoch, da komm ich her") and the second was a popular hymn that was first written in the fourteenth century but performed widely in Protestant lands, "Joseph dearest, Joseph mine" ("Josef, lieber Josef mein").[25] Although the lyrics of the hymns were not sung, it is interesting to note that the imagery of the automaton directly engages them. Luther's hymn pointedly states:

> These are the tokens ye shall mark
> The swaddling clothes and manger dark;
> There shall ye find the young child laid,
> By whom the heavens and Earth are made
>
> Now let us all, with gladsome cheer,
> Follow the Shepherds and draw near
> To see this wondrous gift of God,
> Who hath his own dear son bestowed
>
> Ah, Lord who has created all,
> How hast thou made thee weak and small
> To lie upon the coarse dry grass
> The food of humble ox and ass.[26]

The text of Luther's hymn emphasizes both the preciousness of the most important gift mankind ever received—its savior and son of God—and, paradoxically, the Christ child's humble appearance.

Luther's hymn was widely sung during Christmas and New Year's celebrations throughout the second half of the sixteenth century, and it was often included in Lutheran Weihnachtsspiele.[27] Notably, "Vom Himmel hoch, da komm ich her" was sung in a Weihnachtsspiel titled *Pfund* or *Pfundt*, which Sophie's father Johann Georg I commissioned the playwright Georg

Pondo to write in 1588. Sophie's younger brothers and sisters performed the play on New Year's Day at the Brandenburg court in 1589—the same day Sophie gave Christian I the Christmas Crib Automaton. Three young electoral princes played the roles of the three kings, while three more acted as the shepherds. In Pondo's play, Christ's penniless beginnings and affinity with the common man were accentuated by the actors' use of the northern German dialect Märkisch (which the Margraves and Margravines did not speak).[28] Since her own family performed the play, it is likely that Sophie was familiar with some of its key components: the portrayal of the Christ child and Holy Family as local peasants, and the electoral princes in the guise of the Three Kings. Both figure prominently in the Christmas Crib Automaton.

After the music to Luther's hymn concluded, the Christmas Crib Automaton played the music to "Josef, lieber Josef mein." The lyrics to this hymn, again, address the bereft situation of the King of Kings. The final verse is:

> Little man and God indeed,
> Little and poor, Thou art all we need,
> We will follow where Thou dost lead.[29]

Christian I surely knew the lyrics to these hymns. A significant part of his religious training as a boy consisted of memorizing all of Luther's hymns and many, many others. This suggests that while the music sounded, the lyrics would easily have been recalled, further emphasizing the visual components of the object.[30]

While the hymns played, the kings, each followed by two attendants, paraded around a small, dilapidated structure containing the Virgin, Joseph, the Christ child, an ox, and an ass (figs. 17 and 18). The Star of the East crowned the scene and opened to reveal God the Father, the Holy Ghost, and cherubs, while three more cherubs, frolicking in excitement, descended from the golden sphere. Fixed beneath a suspended God the Father and the Star of the East, who have become fused into one cosmic body, Christ appears God-given, a divinely illuminated, universal sanctified center. It is toward this, the *lux lumens*, that the kings turn. For as the kings passed the manger, they pivoted, faced, and bowed before the Christ child, while the Virgin revealed him by lifting a blanket. Here the kings' conversion, literally their turning toward Christ, ends Christ's condition of concealment. This is the premier *apparatio domini*. Recognition

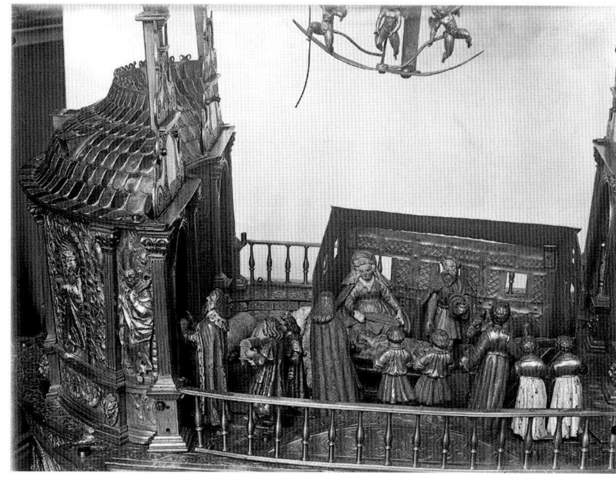

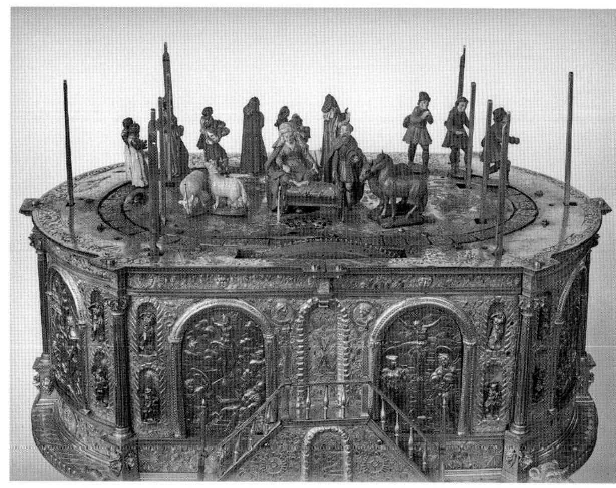

17. Hans Schlottheim, Christmas Crib Automaton (detail), 1588, formerly in the Mathematisch-Physikalischer Salon, Dresden.

18. Hans Schlottheim, Christmas Crib Automaton (detail), 1588, formerly in the Mathematisch-Physikalischer Salon, Dresden.

of Christ's epiphany is in turn registered in the movement of the ox and the ass, who turn their heads toward the cradle that Joseph rocks.

This epiphany requires a material response. After all, Exodus commands, "None shall appear before me empty" (Exodus 23:15). The attendants, who followed each of the kings, carried the gifts of gold, frankincense, and myrrh, which were described in the Gospel of Matthew. Exotic in origin, these commodities not only signaled the kings' foreignness, command of resources, and quasi-diplomatic recognition of Christ's majesty (forecast by Solomon in Psalm 72:11–12, "The Kings of Tarshish and the isles shall bring presents: the Kings of Sheba and Seba shall offer gifts. Yeah, all Kings shall fall down before him: all nations shall serve him") but also modeled visually the practice of salvational giving. At once the gifts convey the kings' recognition of Christ, their faith in him, their sacrifice for him, and their submission to him. It is this act of prestation that sets into motion the human element of the sacred economy of Christianity. It is the moment in which they enter his divine covenant. Do ut des. "I give so that you give."

Once the kings and their entourage passed the Holy Family, they entered a richly adorned semicircular structure and reappeared behind the manger. After the kings passed the back of the manger, they proceeded into yet another semicircular structure that mirrored the first. As soon as they reemerged they repeated their offering. Thus on the Christmas Crib Automaton the narrative of the kings is not given a temporal determination: it is a repetitive, cyclical drama of prestation. The three shepherds of Luke's Gospel followed in the procession. They provide the antithesis to the kings' royal stature, their stances considerably less firm and solemn than those of the travelers from the East, and they serve to draw a distinction between rich and poor, as well as between the court and the country. Notably, the Gospel of Matthew omits the visitation of the shepherds, who according to church fathers figured the local Jews of Bethlehem. With the exception of the shepherds and the fluttering cherubs, the automaton provided the most prominent and renowned features of the Christmas story, as told by Matthew. "And behold the star that [the kings] had seen in the East went before them, until it came and stood over the place where the child was. . . . And entering the house, they found the child with Mary his mother, and falling down they worshipped him. And opening their treasures they offered gifts of gold, frankincense and myrrh" (Matthew 2:9–11). Like Matthew's narrative, the scene atop the Christmas Crib Automaton is sparse. Compared to adoration scenes produced around the same period, the visual field of the automaton appears bare and lacks a distinct setting. The Bethlehem of cavernous, collapsing classical buildings, turbaned merchants trafficking in gems, and exotic animals, like that rendered in so many mid-sixteenth-century Flemish depictions of the Adoration of the Magi, this is not. Its meagerness, its accordance with the Gospel, highlights the main characters of the story. This story is, however, slightly altered in the Christmas Crib Automaton.

As rendered by Schlottheim, all the kings who present gifts to the Christ child were clothed in contemporary courtly garb. But the first king had a curious addition to his collar and red mantle that suggests he was a specific figure rather than a standard type. His mantle and collar were trimmed with the recognizable ermine fur of an elector of the Holy Roman Empire. The same distinctive dress can be seen in Schlottheim's representation of the seven electors on his London Nef, which was made shortly before the Christmas Crib Automaton (fig. 8).[31] The Dresden Kunstkammer contained yet another representation of the electors in their formal garb: the Electors Tankard of 1588 (fig. 19). The tankard depicts the political order of the Holy Roman Empire, with the Holy Roman Emperor seated and flanked on his left and right by the seven electors and their respective insignia.[32] The elector

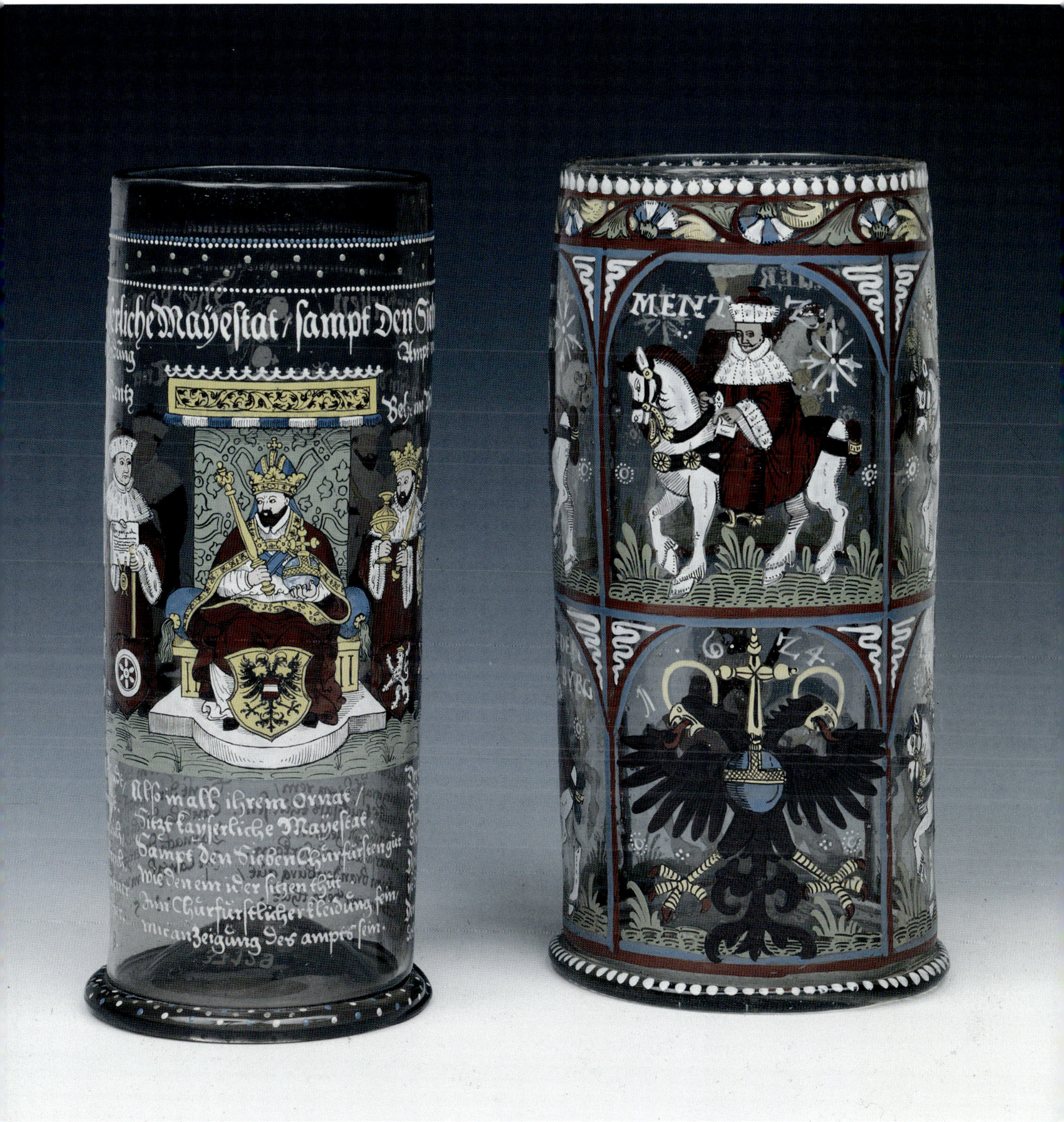

19. Kreibitz School, Electors Tankard, 1588, Grünes Gewölbe, Dresden.

of Saxony stands second from the right between the electors of Brandenburg and the Palatine. All these representations of the elector of Saxony adhere to early modern pictorial conventions of depicting formal attire. Thus the dress of the first king in the Christmas Crib Automaton, despite his lack of crown (Kurhüt), suggests that this figure is an elector, and this formal idiom is a marker for the intended viewer, Christian I.

The Christmas Crib Automaton was a rather flamboyant animation of the Christmas story that participates in the established pan-European courtly tradition of representing a European prince as one of the kings. Holy Roman Emperor Charles IV, for instance, did not just regularly play the role of the first king in festive pageantry: he also arranged for his coronation as king of Italy to take place on the feast of the Epiphany and had himself portrayed as a king from the Christmas story in his residence at Karlštejn. Here, according to Richard Trexler and more recently Joseph Koerner, lurk the ambitions for a universal Christian rule.[33] The same cannot be said about the Christmas Crib Automaton, because it testifies to a world confessionally divided. It exploits the narrative of the Christmas story (by way of Christian's superimposition into it) to convey a particular, local message about epiphany and conversion.

It hardly seems an accident that Christian I was inserted into the Christmas narrative as the first *Christian*. Christian's arrival and literal turn toward Christ are dramatized here; his convertibility is what is staged and animated. The display of the elector in the object, his attire, his pivot, his role as the first king to arrive and respond—these are the elements of the object's religious and political message. Through the representation and animation, his material gift of gold and spiritual gift of faith, Christian's Calvinist identity is reformulated. By way of the object he vows his confessional, his Lutheran, allegiance.

This self-reference on the part of the object's intended recipient was the crucial component of the gift. Both he and other viewers of the automaton could experience the elector as Christian I himself. In a sense it is Christian I who is demonstrating his generosity and largess to the Christ child—an interesting reversal of roles in the context of the New Year's Feast (when, you recall, Sophie of Brandenburg presented this object to Christian), because the patterning and rhythm of the automaton recall ceremonial processions of gift giving that unfolded at princely courts, where the elector or prince would have been the receiver rather than the giver of gifts.[34]

"And Then Rich Men Rushed Forth to Render Their Presents"

In the fifteenth and sixteenth centuries the celebration of the New Year was a favored time for the ritual giving and receiving of gifts. Within the Saxon court and the duchy at large, the tradition even reached the lowest strata of society. Peasants would set a place at the table for a dead loved one, with pieces of bread and a libation. Gypsies throughout the region would bestow the blood and bones of a lamb on fields, in the hopes of a bountiful harvest in the coming year.[35] At court, the giving of gifts during the New Year was conceived of as a reenactment of the three kings bestowing the gifts of gold, frankincense, and myrrh on the Christ child.[36] But instead of the small child receiving the bounty of kingdoms and persons from afar, it was the duke, elector, or prince who was on the receiving end of the ceremonial prestation. Furthermore, the elector did not only receive gifts from visitors and courtiers; he gave them as well.[37] He gave tokens such as barrettes, books, trunks, and clothing to others at court, in addition to which the ruler's visual presence was a privilege and gift in and of itself.

It is not surprising, then, that the New Year was a period when European courts presented themselves at their most resplendent; crowns were donned and regalia displayed.[38] Visitors from nearby and distant lands would come and pay tribute to the prince or duke. Galeazzo Maria Sforza, duke of Milan, wrote of the New Year's festival's ceremonial and representational importance: when the "court [was] frequented at solemn feasts by distinguished and honorable men . . . the prince himself [stood] out more, and the noble and excellent men who [were] in a state of grace and favor in the prince's eyes [grew] in grace and increase[d] in honors."[39] It is clear that one of the key components of increasing one's honor—although only euphemistically referred to here—was showering the duke, prince, or elector with precious tokens, as the travelers from the East did. In keeping with these customs, the Christmas Crib Automaton and representations of courtly ceremonies of gift giving use similar visual strategies of portraying the givers and receivers of gifts.

The "iconography of gift giving" has been stable and consistent since the fifth century BCE.[40] One of the earliest surviving representations of tribute, or gift offering, appears on the monumental bas-reliefs of the northern and eastern staircases of the Apadana, or audience hall, at the ancient ceremonial capital of Persepolis (fig. 20). Designed during the reign of Darius I (522–486/485 BCE), the Apadana was completed

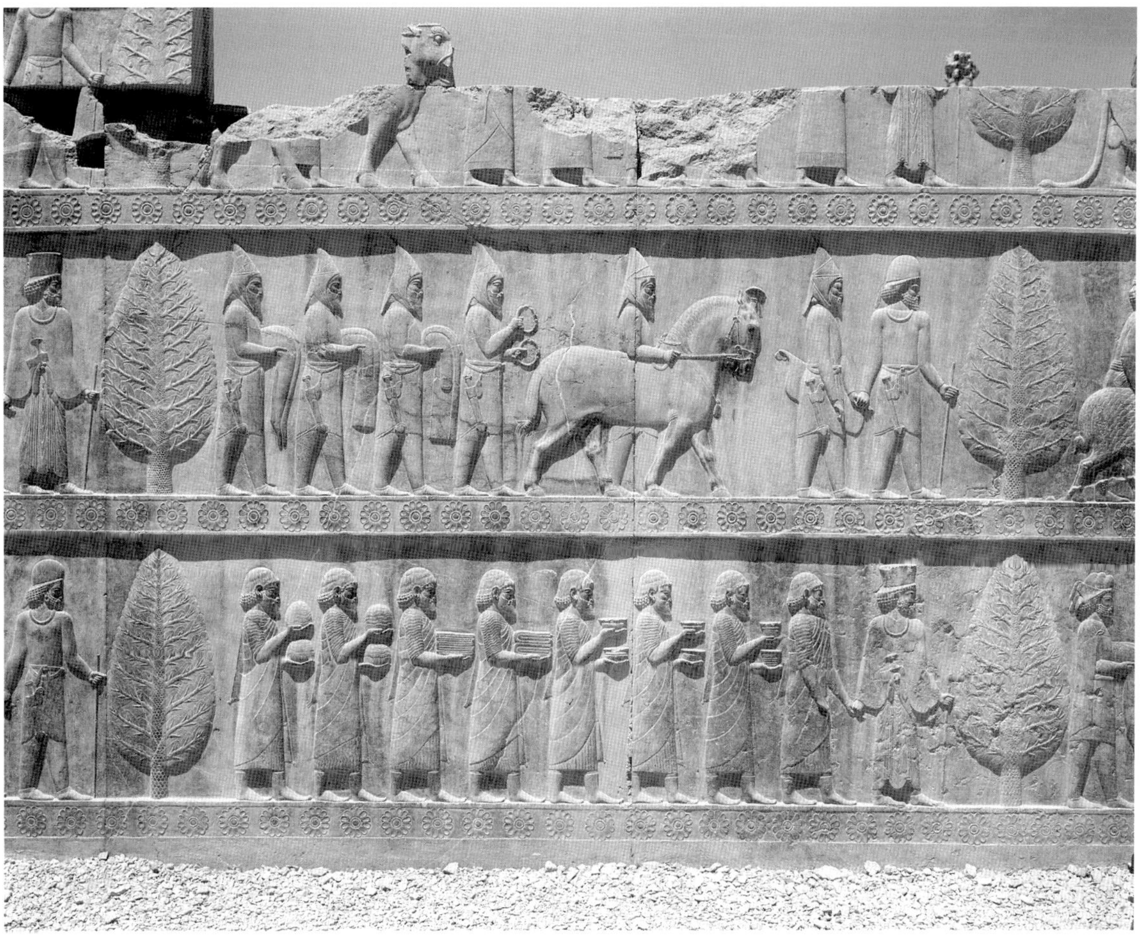

20. Persepolis Eastern Stairway, 522–486/485 BCE.

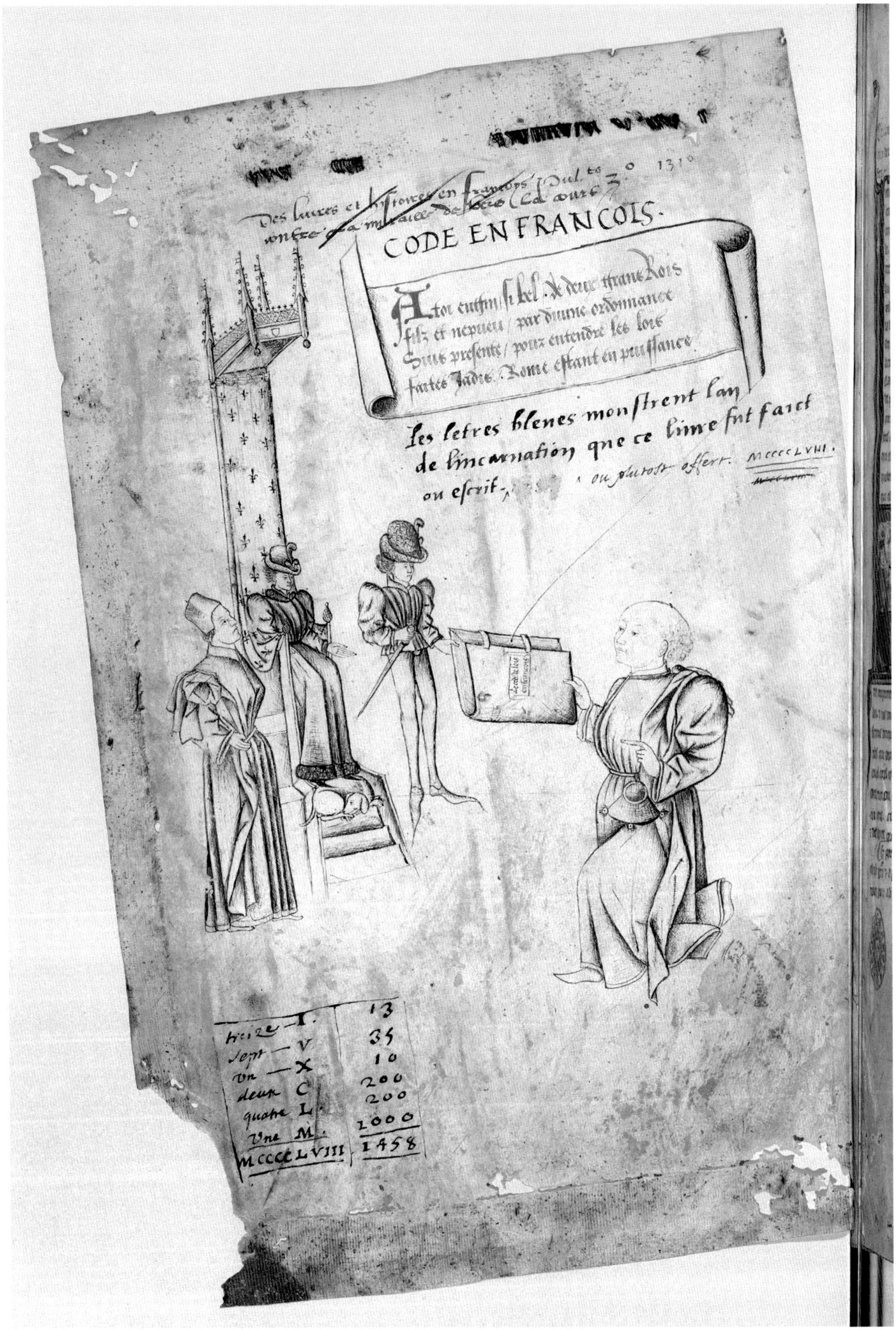
CODE EN FRANCOIS.
Les letres blenes monstrent lan
de lincarnation que ce livre fut faict
ou escrit
ou plustost offert. MCCCCLVIII.
treize — I. 13
sept — V 35
un — X 10
deux C 200
quatre L. 200
Une M. 1000
MCCCCLVIII 1458

under Xerxes I (r. 486/485–465 BCE). It is believed that the reliefs represent the New Year's festivals at Darius I's court.[41] The gift givers are representatives from the twenty-three nations of the Persian Empire. Their gifts, which they simultaneously display to the seated Darius, range from humble vessels, raw materials, and animals to jewels, weapons, elaborate textiles, and gold. The exceptional quality of the reliefs is enhanced by the careful delineation of each gift and of the hair, beards, headdresses, and clothing of the givers. Indeed, when one compares the stance of the tribute bearers in the monumental relief from Persepolis to that of the kings in the Christmas Crib Automaton, their intentional stiffness in contrast to the seated recipients becomes evident. By these means the artists visually demonstrate the dominant and submissive relationships engendered by acts of prestation.

Such an asymmetrical relationship between the giver and receiver of gifts recurs thirteen hundred years later in a ninth-century manuscript illumination of the *Chronicle of John Skylitzes*, now in Madrid (Madrid, Biblioteca Nacional, vitr. 26-2, fol. 75v). In this instance the Byzantine emperor receives priceless documents from the caliph, by way of a messenger. The emperor is depicted on the right, enthroned, and protected beneath a domed structure, possibly a tent. The potentate is in three-quarter view, while the messenger, whose stance suggests genuflection, is depicted in profile—as in the Apadana reliefs and the Christmas Crib Automaton. A further detail, which emphasizes the contrasting status of the emperor and the messenger, is the distinction between the spaces they inhabit. The messenger is clearly standing outside the tent. The white negative space between the body of the messenger and the left corner of the structure highlights the inequality between the emperor and his servant. Similarly, the kings and shepherds of the Christmas Crib Automaton never enter the enclosed space of the Holy Family. Like the messenger, they inhabit the periphery. The only physical link they have to the sacred figures is in the transmission of the gift.

This does not hold true for a third example found in a manuscript of Justinian's *The Institutions* (1458; fig. 21). In this presentation scene Charles, Duke of Orléans, sits enthroned beneath a late Gothic baldachin and before an ornamented curtain bearing the fleur-de-lys. A clerical figure, possibly a bishop, stands to the left of the throne bearing witness to the transaction. The donor or gift giver kneels and passes a book to a finely dressed court official (who, remarkably, dons the same hat as Charles of Orléans) instead of handing it directly to the recipient—yet again signifying the incongruity between the giver and receiver. Despite this discrepancy, this example, like the two previously discussed, presents the fundamental and persistent visual components of representations of gift giving. The giver of gifts is rendered in profile; the offering is clearly displayed to the viewer and the seated recipient. In each of these images, and in the Christmas Crib Automaton, the central action is the transmission of the gift. There is, however, a major disparity between the Christmas Crib Automaton and the examples just discussed. In the automaton the ruler (Christian I) occupies the role of the giver of the gift, and this alteration is indeed significant. Rather than acquiring more wealth, the elector in the Christmas Crib Automaton is giving it away.

The Deed

Martin Luther consistently advocated that the faithful should improve the lot of their neighbors. According to Peter Iver Kaufman, "Luther was convinced that if persons did not wholeheartedly use the goods given to them by God

21. Petrus Schoyff, *Presentation Scene with Duke Charles of Orleans*, from Justinian, *Institutions*, before 1458, Bibliothèque nationale de France, Paris, Ms. fr. 497, fol. Av.

to respond to opportunities for doing good for others, God would deny them the ultimate good of eternal life."[42] At several points in his writings Luther strongly urges against indulging in earthly luxuries and instead stresses the importance of charity, or giving. In his *Sermon on Threefold Righteousness*, which he gave in 1518, Luther unabashedly uses the princes of Saxony to exemplify those whose riches have been bestowed on them by God for their piety: "Therefore Christians, who are to be enriched with eternal good things, are not to be exhorted to righteousness, but rather discouraged from it in favor of a better one. Hence, one is not to rejoice in these things; just as God enriches the princes of Saxony with glory, riches, and pleasure, because they are pious lords. And if these things were not enough, He will bring forth still a mountain of silver and peace in the land will be preserved. But let them see themselves whether this will do them any good for their salvation."[43] Although Saxon princes have been rewarded with seemingly endless silver mines for their devotion, Luther suggests, these riches will not benefit them in their hope for salvation. It is faith first and foremost that allows one to enter paradise. According to Luther, the truly faithful, inspired by their longing for God, give willingly. In *The Freedom of the Christian* (1521) Luther states, "See, therefore, how love and desire for God flow out of faith and out of love flows a spontaneous, willing, joyful life that serves the neighbor for no reward whatsoever."[44] Using the rhetorical imagery of a river, Luther suggests that giving to those in need is a demonstration, an outward manifestation, of one's faith. Ilana Krausman Ben Amos has astutely noted that "Protestantism provided a new language with which to articulate and communicate the benefits of gifts, thus investing acts of giving with religious meaning that would reinforce the impulse to give."[45] Luther also articulates this notion in *The Explanation of the Theses*: "The first and main [good deed] is to help a beggar or one's neighbor in need."[46] And this deed, the paradigm of all pious actions, which displays to the world one's faith, is precisely the deed the kings of the Christmas Crib Automaton perform, as the centripetal force of the motion of the automaton pulls them and their gifts toward Joseph with his tattered hat and his worn clothing, the Virgin on her humble wooden seat, and the Christ child laid to sleep in a barn.

The didactic potential of the Christmas Crib Automaton is in keeping with Luther's considerations on religious imagery. Luther's views on the function of religious art were formed in the first instance as a reaction to the radical iconoclasm promoted by early reformers, such as his Wittenburg faculty colleague Andreas Bodenstein von Karlstadt (1486–1541).[47] Above all, Luther stressed the inevitability of mental visualizations. In his treatise "Against the Heavenly Prophets in the Matter of Images and Sacraments" (1525), he stated, "Whether I will it or not, when I hear of Christ an image of a man hanging from a cross takes form in my heart, just as the reflection of my face naturally appears in the water when I look into it."[48] Second, Luther contended that the work of painters, sculptors, and printers could enhance a layperson's understanding of the Bible. In the same treatise of 1525 he wrote that he wished to "convince the lords as well as the rich to have the entire Bible painted in detail on houses, so that the eyes of everyone could see it."[49] He also saw value in combining mental and visual images and claimed, following Gregory the Great, that people are "more apt to retain the divine stories when taught by picture and parable than merely by words or instruction."[50] For Luther, religious art was an aid to memory; religious images were cues for the faithful to recall significant figures and events, "zum Ansehen, zum Zeugnis, zum Gedächtnis."[51]

Although Luther's views on images were attacked by other radical reformers, such as Huldrych Zwingli (1484–1531), the notions that an art object could serve as a reminder to the faithful and a model of appropriate behavior were adopted by his followers. For instance, in 1526 the Nuremberg reformer Andreas Osiander maintained that the images of saints, which populated many Lutheran churches in Nuremberg, could provide Lutherans with models (*Vorbilder*) of correct conduct: "In the Holy Scriptures there are sufficient and abundant teachings on what we should and should not do. However, we also require, for our feeble wills, good examples and role models of holy and spiritual people. In them we see that God's word teaches us and we should follow it. We do this in order to live immaterial and Christian lives."[52] These examples cited thus far are taken from religious treatises or sermons, which attempted to explicate and legitimize the use of images in the Lutheran church. Remarkably, the same ideas were put forth by Gabriel Kaltemarkt to justify the collection and appreciation of painting and sculpture within a princely Lutheran context.

Almost nothing is known about Gabriel Kaltemarkt. Where and when he was born is a mystery, as is the nature of his education. The little we do know has been deduced from his treatise on collecting, "How a Kunstkammer Should be Formed" ("Bedenken wie eine Kunst-Cammer Aufzurichten seyn möchte"), a text that he dedicated and sent to Christian I in the hopes of being appointed master of the collection (*Kammermeister*) of the Dresden Kunstkammer.[53] In his treatise Kaltemarkt makes clear that he is familiar with many of the important numismatic collections in Europe and has intimate knowledge of the organization and contents of the Dresden Kunstkammer.[54]

Throughout his treatise Kaltemarkt argued that Lutheran rulers needed religion, faithful subjects, money, military equipment, and books, as well as paintings and sculptures, in order to obtain "the best adornment and treasures of a prince."[55] Furthermore, he claimed that art objects were essential to the well-being of a prince. The ruler required books for edification, "but also, as a delight to the eyes and strengthening of memory, likenesses of [the books'] authors and heroes . . . as a living incitement to do good and avoid evil."[56] These beliefs regarding the ennobling function of art objects most likely governed how the Christmas Crib Automaton was endowed with religious and political agency. Within a Lutheran framework, the three kings on the automaton were reminders or signals for Christian I that reinforced the proper actions and duties of a devout Lutheran.

If arranging marriages for strategic purposes was intended to solidify the relationships between Protestant courts, it also opened the door for disagreement. Toward the end of his reign Elector August I took great pains to foster a smooth transition of power. He attempted to flush out the Calvinists at court and aligned himself with the orthodox Lutheran court of Brandenburg. And finally, he brought his heir, Christian I, into the political proceedings before his ascent to the office of elector. In light of the fragile political situation surrounding the early years of Christian's reign and Sophie of Brandenburg's strong Lutheran stance, a religious and political interpretation of the Christmas Crib Automaton seems fitting. The tension between husband and wife, one a crypto-Calvinist and the other an unwavering Lutheran, embodies the anxiety of late sixteenth-century confessional politics in the Holy Roman Empire.

Although there is no known surviving correspondence between Sophie of Brandenburg and Hans Schlottheim that describes or discusses the commission of the automaton, it seems appropriate that Sophie would have commissioned such an object for her religiously dithering husband as an exemplification of proper religious and therefore political duties. The gift of the Lüneburger Spiegel (fig. 16; ca. 1587) to her son Christian II—Christian I's successor—on the eve of his becoming an elector demonstrates that she was prone to giving religious and politically loaded gifts to her immediate family.

The *Fürstenspiegel*

At the time of his father's unexpected death, in 1591, Christian II was only eight years old. Owing to his young age, his grandfather Johann Georg I of Brandenburg and the young prince's Ernestine cousin Duke Friedrich Wilhelm I of Saxony-Weimar-Altenburg stood in as regents of the duchy until 1601, when Christian II turned eighteen and was able to take on the responsibilities of an elector. The ten years between Christian I's death and Christian II's ascension were dominated by yet a second Calvinist blitz at the Saxon court. Immediately after Christian I's demise, many of his consorts were thrown into prison for their Calvinist beliefs. Most notably, Nikolaus Krell was imprisoned and executed just days before Christian II took office.[57] His execution was very likely a symbolic act, to demonstrate that the Saxon court had cleansed itself of any residual Calvinist leanings. Therefore, as with the Christmas Crib Automaton, Sophie of Brandenburg's gift of the Lüneburger Spiegel occurred at a weighty political moment.

In 1600 Sophie requested that the Dresden Rentkammer dispense 1,450 gulden to Johann Schlowen of Lüneburg to pay "for a large mirror fitted with 1,000 pieces of pounded and gilded silver. It also shows all of the coats of arms of the Holy Roman Empire and is adorned with many Bohemian gems."[58] In addition to its sparkling gems and coats of arms, the Lüneburger Spiegel is densely populated with representations of the prophet Daniel's analysis of Nebuchadnezzar's dream and allegories of good and bad government.[59] In this mirror allusions to a Lutheran world order abound, and the visual and verbal programs are particularly well suited to its recipient, the newly instated elector of Saxony.

The mirror was designed and crafted by two goldsmiths, Luleff Meyer and Dirich Utermarke, between 1587 and 1592.[60] Meyer was a master from Lüneburg, while Utermarke was an independent master (*Freikünstler*) working in Hamburg.[61] Scholars have disagreed over the original patron of the Lüneburger Spiegel. Hans Schröder, the first and thus far only art historian to devote a monograph to the mirror, assumed that Sophie of Brandenburg commissioned it in 1587. Schröder based his hypothesis on the Dresden Kunstkammer inventory of 1610, which includes the following description of the mirror: "One large gilded silver mirror adorned with Bohemian gems, on which appear Daniel's entire prophecy of the four monarchies and the coats of arms of the kingdoms, duchies and provinces of the Roman Empire. It is from the elector's widow, who bought it from a man in Lüneburg. Additionally, the writing on the glass is executed in enamel."[62] Following Schröder, J. F. Hayward also claimed that Sophie commissioned the object. However, Dirk Syndram has argued that Sophie did not commission the mirror but merely purchased it from Johann Schlowen of Lüneburg.[63] Syndram's argument has most recently been expanded upon by Susan Tipton on the basis of documents that record the construction and decoration of the Lüneburger Rathaus. Tipton found that the Lüneburger Spiegel was most likely commissioned for the Rathaus but that the funds needed to purchase it were not available. The Lüneburger Spiegel was consequently placed on the open market and purchased by Sophie eight years after its completion, in 1600.[64] Although Sophie may not have played a role in designing the object, the fact remains that she bought it with the intent of giving it to her son Christian II.

The shape of the mirror's frame is derived from monumental epitaphs in northern German churches, which commemorated the nobility or wealthy merchants.[65] The epitaph for Heinrich von Schönberg (ca. 1575), which resided in the Dresden Frauenkirche until the church was destroyed in 1945, is a key example of the form (fig. 22). The frame of the mirror is a densely concentrated space incorporating a wide range of ornamentation: swags of drapery, *Rollwerk*, oval bosses, flutes, and trophies of arms.[66] In addition to its swarming decoration, the object includes figures cast in the round, relief work, and engraved detail. All the figures on the top half of the mirror are derived from 2:27–45 of the Book of Daniel, in which the Hebrew prophet divines and analyzes the dream of Nebuchadnezzar, the Babylonian king and Daniel's captor. In his dream Nebuchadnezzar saw a colossal statue whose head was made of gold, chest and arms of silver, belly of brass, legs of iron, and feet of clay.

In the vision, a singular stone cast into the air destroyed the massive statue. After being struck by the stone, the statue turned to dust and blew away "like the chaff of summer threshing floors" (Daniel 2:35). Yet the stone remained and became a mountain, which covered the entire earth and stood for eternity. The prophet Daniel interpreted Nebuchadnezzar's dream as a nebulous timetable that concluded with the domination of the Kingdom of Heaven. Daniel explained that the head of gold signified the kingdom of Babylonia, while the chest and arms of silver and belly of brass foretold of two kingdoms that would follow but be inferior to Nebuchadnezzar's empire.[67] And finally:

> There shall be a fourth kingdom, strong as iron; as iron shatters and destroys all things, it shall break and shatter the whole earth. As in your vision, the feet and toes were part potter's clay and part iron, it shall be a divided kingdom. Its core shall be partly of iron just as you saw iron mixed with common clay, as the toes were part iron and part clay, the kingdom shall be partly strong and partly brittle. As in your vision, the iron was mixed with common clay, so shall men mix each other by intermarriage, but such alliances will not be stable: iron does not mix with clay. In the period of those kings the God of heaven will establish a kingdom which shall never be destroyed; that kingdom shall never pass to another people; it

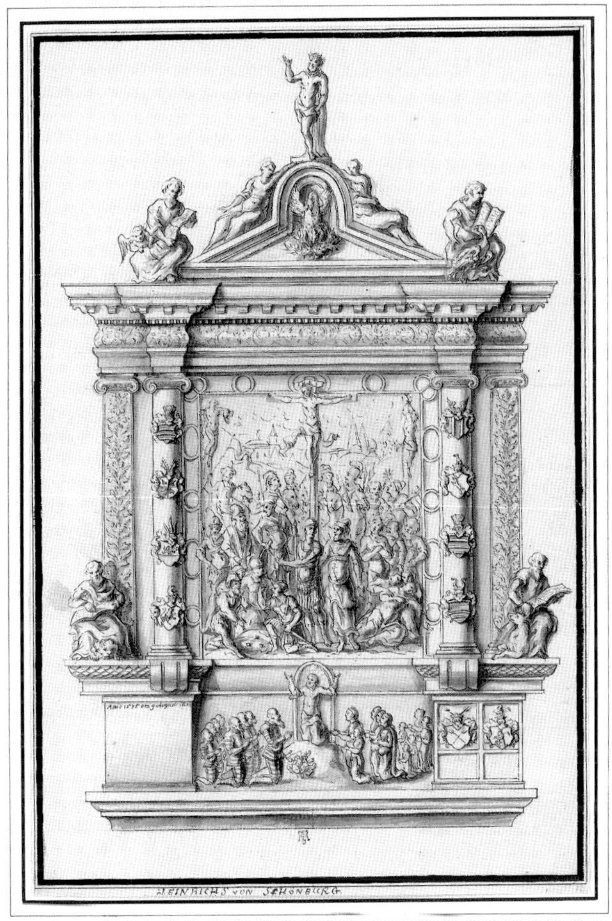

22. *Epitaph for Heinrich von Schönberg*, in Gottfried Michaelis's *Dresdenisch Inscriptiones und Epitaphia, welche auf denen . . . in und ausser der Kircher zu unser Lieben Frauen . . . zu finden*, 1714, Kupferstichkabinett, Dresden.

shall shatter and make an end of all these kingdoms, while it shall itself endure forever. (Daniel 2:40–44)

It is significant that half of the decoration of the Lüneburger Spiegel is dedicated to the Book of Daniel, because the book and the prophet himself were treated repeatedly by Luther in his sermons and writings that addressed the fate of the Holy Roman Empire and its leaders.[68] Luther capitalized on Daniel's apocalyptic analysis and intimated that the final divided kingdom was none other than the confessionally split Holy Roman Empire. In his "Preface to the Prophet Daniel" ("Vorrede über den Propheten Daniel") Luther declared, "The Holy Roman Empire will remain until Judgment Day. The kings and popes are powerless. Daniel does not deny this and heretofore the record has attested to it."[69] This apocalyptic rhetoric was picked up by Luther's students. Johann Mathesius (1504–1565), in his publication of his *Bergpostilla* (1562), hailed Luther's exegesis of the Book of Daniel. Visual renderings of Luther's interpretation of the colossal statue began around 1550, most notably with a woodcut by Hans Brosamer.[70]

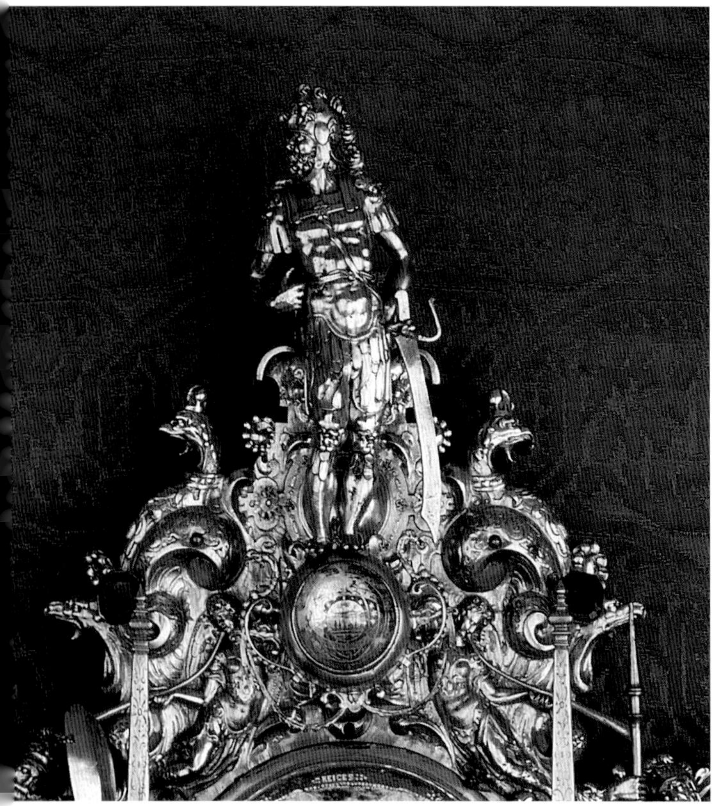

23. Luleff Meyer and Dirich Utermarke, Lüneburger Spiegel (detail), 1587–1592, Grünes Gewölbe, Dresden.

The visual program of the Lüneburger Spiegel deploys Luther's exegesis of the Book of Daniel while imbuing it with political significance through the mingling of figures from Nebuchadnezzar's dream with allegories of good and bad government and heraldic devices. The mirror is divided into five registers, organized along a vertical axis. Soaring above the tangled ornamentation is the Statua Danielis, which appeared to Nebuchadnezzar.[71] This figure, dressed in a Roman cuirass, stands with one foot firmly planted upon a painted image of the city of Lüneburg. His sword, which echoes the curve of his lower body, hangs precariously from his left hip, while his immense and sinewy hands rest against his torso. Although the figure's body faces frontally, his head is turned to the right. The statue's distinctive dress, particularly his leonine padded knees, identifies him (fig. 23). The same formal attire adorns the body of the statue in Lorenz Faust's *Anatomia Statuae Danielis* (1586; fig. 24). The second register is dominated by the painted medallion of the black double-headed eagle of the Holy Roman Empire, whose spread wings form the backdrop for the fifty-six coats of arms of the constituent bodies of the empire, while the crucified Christ is placed centrally on the bird's breast. Flanking the medallion are two warriors on rampant horses. The horses' contorted heads draw the viewer's eye back to the medallion and the smaller figures of Neptune and sea nymphs, who stand before the Corpus Christi. The second register is punctuated by a row of twelve more painted medallions, each containing a coat of arms from the Protestant lands in the empire.

The central and largest register includes the mirror, which is protected behind a large rectangular panel depicting the Earth trapped in the net of Vice and crushing the back of a personification of wickedness.[72] To the right of the Earth stands Mars, whose warring proclivities are symbolized by the burned and pillaged city strategically placed above his head in the middle ground. Peace stands to the left of the Earth, her long, attenuated arm drawing the eye to the undisturbed landscape, filled with mountains and quaint buildings. The center of the panel is dominated by two angels. One is perched on top of the Earth, while the second floats above, sounding the trumpet of the apocalypse and heralding the coming of Judgment and the dawn of the Age of Christ. The central panel is bordered by two warriors set in niches, while two other warriors rest on the extravagant volutes and display the trophies of war. These four warriors represent the four fallen kingdoms that Daniel refers to in his analysis of Nebuchadnezzar's dream.[73] They bridge the first and third register because each kingdom, according to Daniel, was embodied by the Statua Danielis. Like the second register, the third is punctuated by yet another row of painted medallions, which present twelve more coats of arms of the Protestant lands of the Holy Roman Empire. An allegory of good government is situated below the medallions. This conglomeration of figures shifts attention away from the constituent bodies of the Holy Roman Empire and the apocalyptic visions of the top half of the mirror and toward the purpose of government and the virtues of a prince.

The allegory of good government consists of hierarchically organized personifications of moral and civic concepts. In the large niche Justice sits, elevated in the center, while Prudentia stands to her left and Caritas, holding a writhing baby, stands to Justice's right. Pax and Res Publica grasp hands at the feet of Justice.[74] The allusion to the necessity for the state of maintaining peace is amplified by the chains attached to Pax's and Res Publica's wrists, which terminate around the wrists of Justice. Consequently the chains binding the three figures take the place of the balance in traditional representations of Justice. Instead, Justice's entire body functions as the balance, and her domination over Pax and Res Publica clearly expresses the necessity to "have ordered all things with measure, number, and weight" (Wisdom 11:20) in the maintenance of *tranquillitas, civium*.[75] This register is separated from the fifth and final register by a third set of painted medallions. Unlike the previous heraldic devices, these medallions announce the princely virtues: *caritas, iustitia, patientia, spes, fortitvdo, and temperantia*. The verbal and pictorial programs reinforce and engage one another, with the recitation of

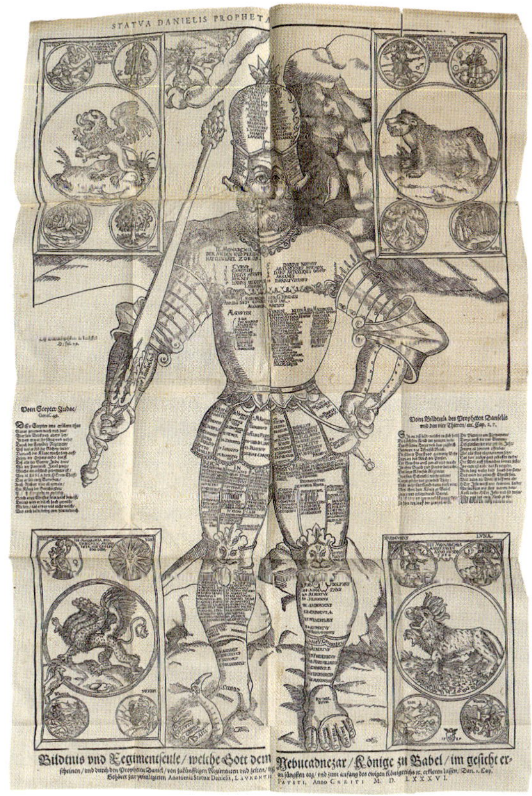

24. Lorenz Faust, *Anatomia Statua Danielis*, from Lorenz Faust's *Anatomia Statuae Danielis*, 1586.

the princely virtues directly beneath the allegory of good government demonstrating that the allegory was intended for the eyes of a princely ruler.

The judgment of Paris is rendered below the princely virtues. He stands in the center while Apollo hands him the golden apple to give to either Aphrodite or Helen of Troy. The mirror presents the moment before Paris's fatal decision, which led to the destruction of his kingdom at the hands of the Greeks. In this way the final register functions as a commentary on the importance of good judgment for the sake of one's kingdoms. The judgment of Paris is intended to correspond, antithetically, to the princely virtues and allegory of good government. Thus does the Lüneburger Spiegel bring together terrifically complex visual and verbal programs that highlight the duties and virtues of a ruler of Protestant lands, against the backdrop of the end of days. In Christian II's possession, the mirror reminded the Saxon ruler of his Lutheran faith, his allegiance to the Protestant lands, his duty to maintain a proper and peaceful government, the princely code of conduct, the consequences of poor decisions, and his prominent role in an empire that will, according to Luther, be the last of all empires.

This final aspect of the object becomes even more prominent when we consider that a vast amount of ornamentation functions as a frame for a mirror, a surface that has a very particular ocular dynamic. When Christian II gazed into the mirror, its reflective surface incorporated his reflection, his own image, into the object, placing him at the center of an overtly Lutheran political allegory.[76] Therefore, the inclusion of Christian II's image in the Lüneburger Spiegel functions in ways similar to the animation of the elector in the Christmas Crib Automaton. Both objects make their intended recipients visible. Both play on the pleasure of marking the self as remarkable. Both enable the rulers to see themselves in their respective gifts as virtuous Lutheran rulers. And each was given by Sophie of Brandenburg to its recipient at a politically momentous time. The similarities, however, end there.

When Christian II gazed into the mirror, its surface created a reflection, and it permitted his self-image to be seamlessly incorporated into the object. As if unmediated, Christian II experienced the appearance, the epiphany, of his own face. Consciously or unconsciously he would have registered the synchronization between his actions and movements—the blink of an eye, the furl of the brow, the twist of the mouth, the wrinkle of the nose—and that of his counter-image. The correspondence between viewer and mirror image, as well as the proliferation of imagery that framed and staged it, was undoubtedly intended to arouse Christian II's self-regard. The Lüneburger Spiegel was not manufactured and then purchased to create a casual reflection. But the correspondence between the viewer and his bent-back image contributes to the political significance of the object in two further and interrelated ways. The rhetoric of the mirror's presentation was one of truthfulness, implying that there existed a harmony between reality and reflection and that the harmony was in the present tense: the image upholds the status quo. Although surrounded by political and religious allegorical imagery, Christian II saw himself as he was at the moment when he gazed into the mirror's surface, and that self-image is celebrated by the pictorial program of the mirror.

The Christmas Crib Automaton does just the opposite. It portrays an ideal. The automaton manifests a disjunction between Christian I and his representation in the object, not because there exists no direct congruity between reality and representation but because the automaton reformulates Christian I's identity; he is, after all, represented as the first of the three kings of the Christmas story—the first Christian. And I argue that in Christian I's potential recognition of this reformulation is where the message of the Christmas Crib Automaton becomes apparent. The message is this: Christian must, like the king from the East, convert; like the king on the automaton, he must turn toward Christ and give the gift of his faith. In doing so he, like the first Christian, will cease being an outsider and become part of the one and true confession again.

Of course it will never be known whether Christian I gazed upon the Christmas Crib Automaton with delight, or whether he recognized a version of himself recognizing Christ in the object and understood it to be an ideal model of his position at a confessionally torn court and duchy. Close examination of the object has revealed not only that its imagery speaks to the customs and rituals pertaining to the time of year when it was given, but also that the complicated issues of Christian's Calvinist leanings underpin both the music and the imagery of the automaton. While Sophie of Brandenburg's motivation for commissioning the object remains a matter of speculation, we can be certain that she employed an analogous strategy of self-recognition in her gift of the Lüneburger Spiegel to Christian II. If representing an elector in an alternative guise could promote desired behavior through Lutheran ideas about the function of art objects, then mirroring the Lutheran ruler, transforming him into the model, makes clear that his achievement has been fulfilled.

Four

A Figure of Speech

Twenty-seven years after Sophie of Brandenburg presented to Christian I of Saxony Hans Schlottheim's Christmas Crib Automaton, Ferdinand Wittelsbach (1577–1650), the archbishop of Cologne, provided the China Mission of the Society of Jesus with a mechanized nativity scene. Nicolas Trigault (1577–1628), the Flemish Jesuit who accepted the object, and who had transported it to Macau when he traveled there in 1619, described the automaton in a widely published letter to Pope Paul V (1552–1621); it included the three Magi in a procession before the Christ child asleep in a cradle, the Virgin and Joseph rocking the cradle, an ass and an ox turning their heads toward the Christ child, God the father making benediction, and angels continuously descending and ascending.[1] Ferdinand was not the sole member of the Wittelsbach family to endow the China Mission. His brother, Wilhelm V of Bavaria (1548–1626), pledged to Trigault to donate annually eight hundred gulden to the mission—a promise kept by Bavarian dukes for many decades to follow.[2] Wilhelm furnished Trigault with several reliquaries, a clock in the form of a cross, and a silver clock adorned with figures of the Virgin, the Christ child, and John the Baptist. Maximilian I (1573–1651), Wilhelm's son and the duke of Bavaria, gave Trigault a richly decorated cabinet containing an array of mechanical instruments, among them an odometer, clocks (too numerous, according to Trigault, to count), and an astrolabe. Perhaps the most impressive object came from Maximilian's wife, Elisabeth Renata of Lorraine (1574–1635). According to Trigault, she added "a centaur that runs across a table by itself, shoots an arrow with such force that [the arrow] could even be driven into the wood close-by and be fixed; it strikes the hours with its hoof and fiercely tosses its head at each individual moment and displays some other actions that escaped my notice."[3] Trigault included this vivid account of the centaur automaton in his published letter intended for the pope and potential financiers of the Jesuit mission—that is, Catholic rulers and intelligentsia. Trigault's description of the centaur is underwritten by the fantasy, powerful in Counter-Reformation Europe, that clockwork objects would help bring exotic foreigners, particularly rulers, into the fold of the church.[4]

The Wittelsbach family deployed similar mechanisms to bolster the church in other instances as well. Few automata are as complexly located in early modern visual and material culture as is the only automaton that the Wittelsbachs prominently displayed in the ducal Kunstkammer in Munich: the Verkehrte Welt Automaton, or automaton of the world upside down (fig. 25).[5] In this oppositional world a golden ape rolls its eyes and beats a pulpit before three white enameled deer—two stags, one doe—atop an iridescent green hill in a sylvan landscape. Made sometime between 1560 and 1570 in Augsburg, the object is one of the earli-

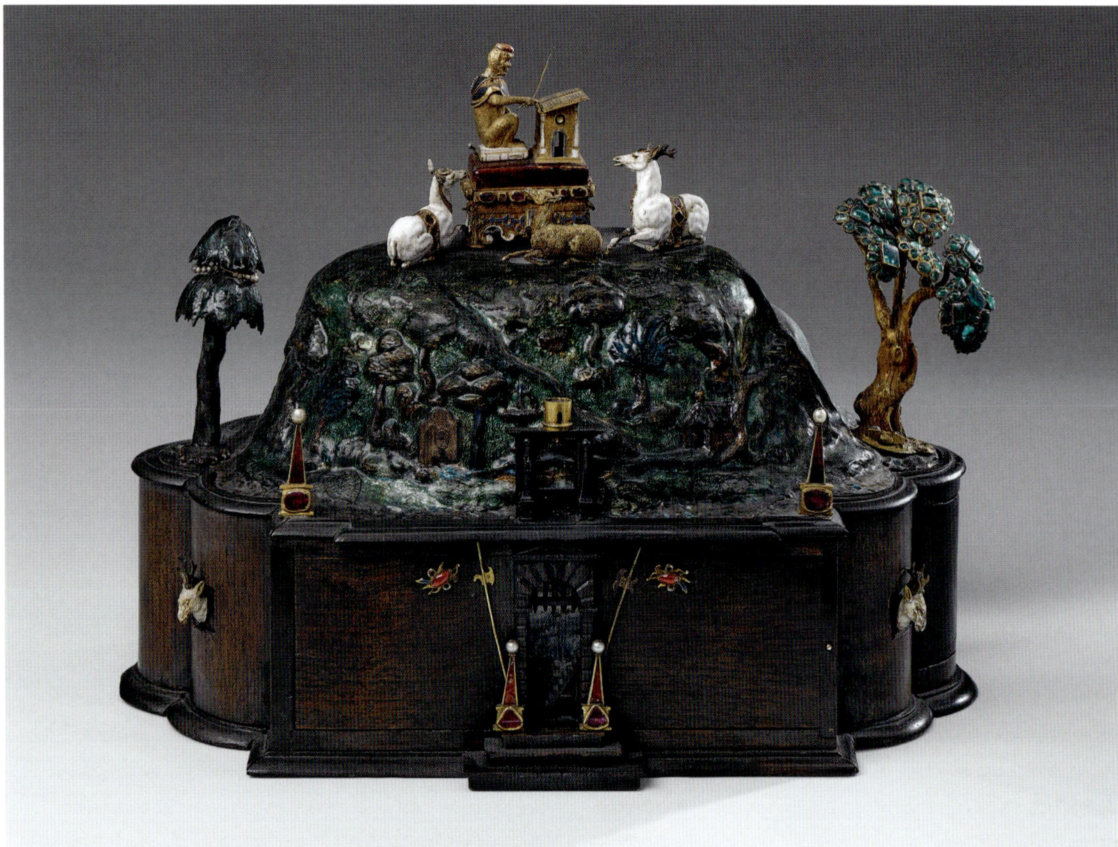

25. Verkehrte Welt Automaton, 1560–1570, Residenz Schatzkammer, Munich.

est extant sixteenth-century German automata as well as one of the most diminutive (it stands only 14.2 cm tall). The precious materials—rubies, pearls, diamonds, emeralds, silver, gold, enamel, and ebony—suggest that the automaton was valued for its intrinsic worth, and that it suited the well-documented extravagant tastes of the Wittlelsbachs, who, like most early modern princes, enjoyed expensive and brilliant things, and who, like many, were admonished by their council for doing so.[6] Bejeweled objects were numerous in the Kunstkammer in Munich, among them the *Statuette of Saint George*, with rubies, emeralds, and opals; Hans Reimer's *Sapphire Cup*; a Ceylonese ivory chest adorned with rubies; and celestial and terrestrial globes ornamented with dangling pearls.[7] The Kunstkammer even housed a lavishly illustrated manuscript containing 110 miniatures by the court painter Hans Mielich that pictorially inventoried Duke Albrecht V's and Princess Anna of Austria's gems and jewelry collection.[8] To limit an account of the Verkehrte Welt Automaton to its place among the Wittelsbachs' bejeweled collectibles, though, would be to ignore how its imagery and animation spoke to other local preoccupations, namely the exculpation of the Catholic Church and the condemnation of the Protestant reform movement. The Verkehrte Welt Automaton visualizes an acute argument, dear to the Wittelsbachs, against one of the central tenets of Martin Luther's reformation.

Pietas Bavarica

By the second half of the sixteenth century, when the Vekerhte Welt Automaton was crafted, the court in Munich had established itself as distinctly confessional. In Bavaria the persecution, prosecution, and expulsion of Protestants was sanctioned from above, for the first time in German-speaking lands.[9] On Ash Wednesday 1522 the conjoint rulers of the territory, Dukes Ludwig X (1495–1545) and Wilhelm IV (1493–1550), issued an ordinance decreeing that followers of Luther were forbidden within their realm.[10] Enforcement of the ordinance began immediately, and all those rumored to have Protestant sympathies were subject to expulsion. In 1522 Wolf Russ, a chaplain in Neuötting, was forced to flee to Ulm after the archbishop of Salzburg, Matthäus Lang von Wellenburg (1469–1540), learned that Russ had publicly claimed the pilgrimage shrine of Our Lady at Altötting to be a *heidnischen Kult*.[11] Reformers were accused of heresy, put on trial, deprived of their property, dismissed from their positions, and publicly humiliated. Just one year after the Russ affair, a young professor at the University of Ingolstadt, Arsacius Seehofer, was arrested for his proto-Philippist lectures on Paul's Epistles, forced publicly to recant seventeen articles of evangelical teaching, and imprisoned in a Benedictine monastery in Ettal.[12] Responding to the public abasement and seizure of Seehofer, Argula von Grumbach (née von Stauf; 1492–1554), a noblewoman from Ingolstadt, published a letter addressed to the Bavarian dukes and the university, in which she criticized the state's and university's treatment of the professor. Von Grumbach's command of Scripture (she cites and discusses more than eighty passages in the letter) was praised by Luther himself in *Wider das blind und toll Verdammnis der seibzehn Artikel von der elenden Schändlichen Universität zu Ingolstadt ausgangen* (1524). Shortly thereafter she became widely known among Protestants as the *neuen Judith*. Anxious over von Grumbach's popularity, Ludwig X and Wilhelm IV, aided by the conservative theologian (and later inquisitor) Johannes Eck (1486–1543), instigated a smear campaign, in which von Grumbach was repeatedly referred to as the *schändlich Weib*.[13]

The purge of the Protestants continued into the reign of Duke Albrecht V, alongside a new initiative to disable reformers' ability to communicate among themselves.[14] In 1558 Albrecht V issued an ordinance that decreed: "Henceforth no bookseller, whoever he may be, resident or alien, may secretly or openly peddle or seek books, be they in Latin or German, that deal with theological matters, in which the Holy Scriptures are discussed . . . and interpreted, or [books] that defend this or that teaching and confession; likewise no books or hymnals . . . [are to be] brought into the land except for those printed in the following cities and country: Munich, Ingolstadt, Dillengen, Mainz, Cologne, Freiburg in Breisgau, Innsbruck, Paris, Lyon, Venice, Rome, Florence, Bologna, Antwerp, Louvain, and Spain."[15] Not only did Albrecht attempt to thwart the circulation of Protestant texts in Bavaria; he also policed the flow of persons. Merchants in Munich were not permitted to travel to confessionally mixed cities such as Augsburg, Nuremberg, and Regensburg.[16] In addition, anyone caught crossing Bavarian borders to attend Protestant services was fined fifty to one hundred florins.[17] Nor were Bavarian students granted permission to attend Protestant universities such as Wittenberg and Leipzig in Saxony, and when students did leave Bavaria to study they were required to present proof of their matriculation at Catholic institutions.[18]

By the time Wilhelm V guaranteed his benefaction to the Jesuit mission in 1616, the Wittelsbachs had maintained an intimate relationship with the Society of Jesus—by far "the most active, most visible, and the most successful agent of the Roman Church and Catholicism"—for more than half a century.[19] Veiled rumors of a problematic alliance between the Wittelsbachs and the Jesuits haunted the legacy of Wilhelm's father, Albrecht V. In the spring of 1580 the Lutheran elector of Saxony, August I, wrote a letter to Wilhelm V asking him to confirm the following piece of gossip: that a kidney stone in the shape of a Jesuit's head had been extracted from Albrecht V's body during an autopsy. August included in the letter a drawing he had received of the kidney stone as proof of the prodigy. The broad and indelicate innuendo was vivid. The Jesuits had maintained authority over Albrecht V and exerted so much power over the direction of his policies, particularly constraints on practicing the Lutheran confession, that their image was indelibly impressed onto his body. Wilhelm responded with haste and countered that the drawing was false and the story a slanderous product of the confessional tension in the empire. Wilhelm also included a drawing of the kidney stone on which the head of a Jesuit did not appear.[20]

If the letter was not a direct consequence of Protestant and Catholic strain, it was surely inspired by the house of Wittelsbach's commitment to a particular commixture of politics and piety. *Pietas bavarica*—a dynastic myth claiming that God had chosen the House of Wittelsbach to be the defender of Catholicism—permeated ducal political statements, correspondence, ambassadorial reports, legislation, the education of Wittelsbach children, the establishment of universities in their realm, the duke's relationship with lower nobility, marriage alliances, and the reputation of the family in the Holy Roman Empire and abroad.[21] Throughout most of the sixteenth century and well into the seventeenth, in their patronage of pictorial programs for Jesuit churches, in portraits publicizing their image, and in the many objects they collected and displayed in the renowned ducal Kunstkammer in the Munich Residenz, the dukes of Bavaria supervised a wide spectrum of art production that exalted the Catholic Church and cast aspersions on the Protestant faith.[22] Objects and images that lauded the Wittelsbachs' alliance with the Catholic Church populated the ducal Kunstkammer. Some, like an elaborately illustrated edition of Giovanni Battista Cavalieri's *Ecclesiae Militantis Triumphi* (1583) and numerous portraits of cardinals, bishops, and popes, announced that the members of the House of Wittelsbach considered themselves not just, in Wilhelm's own words, "worldly princes, but prelates, wardens of the church."[23] Other objects conveyed the dukes' piety: a wooden model of the holy sepulcher; a band of blue, white, and red cloth used to measure the length of the Virgin's body; a rosary made of the earth from which Adam was molded; and three pieces of earth from Palestine bearing representations of Christ and the paschal lamb.[24] Bread, porridge, and flour made from grain that miraculously fell from the heavens in Braunau, Pfarrkirchen, Mattighofen, Burghausen, and, last but not least, Wittelsbach near Asbach in 1570 were evidence that Bavaria was in God's good graces.[25] While objects such as these conveyed that the Wittelsbachs and the church were united in a common purpose, other objects and images overtly and obliquely condemned and satirized the Protestant confession, such as an assortment of Counter-Reformation propaganda prints that were bound together.[26]

The most novel and resplendent of these anti-Protestant objects was the Verkehrte Welt Automaton. Its ensemble of a hilltop, monkey, audience of deer, and trees rests on an ebony base containing four drawers. Each drawer houses a miniature golden hunter's whistle, and the drawer pulls are small, white, enameled deer heads.[27] The front of the ebony base contains an arched portal. There are three steps at the foot of the portal, and the top is adorned with a pediment. Inside the portal a descending drawbridge partially obstructs a minutely rendered landscape on parchment, suggesting that the base is a fortress in miniature. The reference to defensive architecture in the base is underscored by the two gold battle-axes flanking the portal. These instruments of war are framed by two ruby obelisks, each crowned with a single pearl, while four more ruby obelisks are placed at the four corners of the base. A small ivory Roman soldier stands in a niche on the back of the burnished fortress. His whiteness, like that of the enameled deer heads, contrasts strikingly with the black south Asian wood of the base.

Two twisting golden trees sprout from the right and left sides of the enameled mound atop the base. The leaves of these trees, one deciduous and the other coniferous, are denoted by nothing less than translucent emeralds. The hill, which was crafted in the difficult technique of *émail en ronde bosse*, contains embossed trees and two small huts (fig. 25).[28] Reigning over this topsy-turvy world is a golden ape, mammoth relative to the scale of the object, mounted on a red, enameled platform behind a diamond-studded and repoussé enameled golden pulpit (fig. 26).[29] The open songbook before the ape—a hymnal, perhaps—suggests that the animal is in the midst of a service. That the ape is preaching is strongly suggested by the gem-studded pulpit, in combination with the creature's mechanical movement; powered by a clockwork mechanism encased in the enameled landscape, the ape turns its head, rolls its eyes, and raises and lowers its right arm, repeatedly and emphatically beating his radiant, jeweled lectern. Before the ape lie the three enameled deer. Their heads are raised, their ears pricked up, and their gazes directed toward the ape. Although these animals seem at home within the landscape, their juxtaposition with a rhythmically gesturing, preaching ape at a pulpit is incongruous—and iconographically novel.[30] Indeed, the relationship between the ape and the deer is marked by alienation. The bewildering coexistence of the ape and the deer, the ape's mindlessly repetitive gestures, and even the silence of the object make the Verkehrte Welt Automaton a telling commentary on the act of preaching. The combination of gems, gemlike colored enamels, gold, and silver—all materials that refract and seem to emit light—enhances the object's bewildering imagery and movement. What is strangest about this automaton, however, is not just the bizarre coupling of an ape and deer, but its attempt to convey that the ape is communicating with the deer. This is a world turned upside down and inside out, a world that "neither is, nor can be, nor has been."[31]

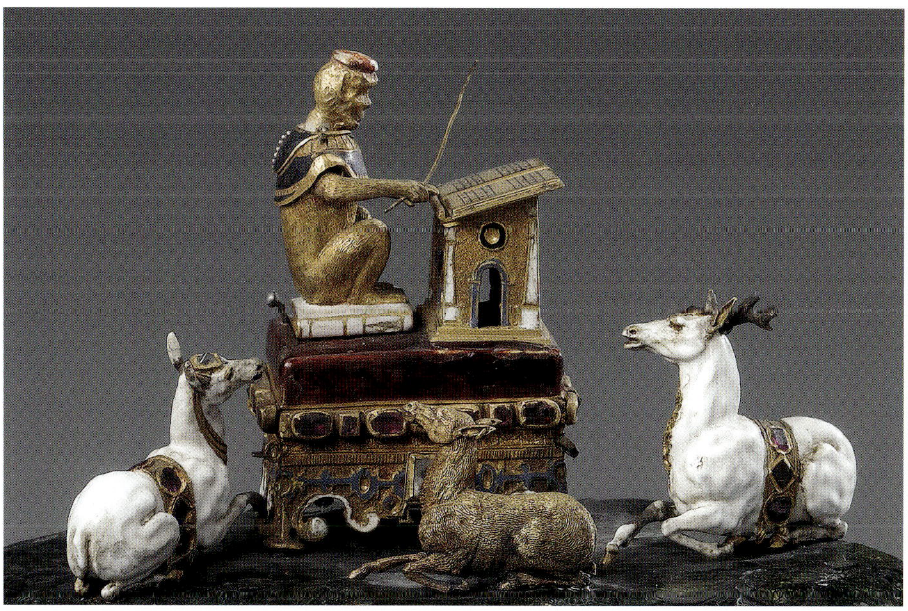

26. Verkehrte Welt Automaton (detail), 1560–1570, Residenz Schatzkammer, Munich.

The Mimic, the Deceiver, and the Fool

Writing in 1611 during a visit to the Munich court, the famed merchant Philipp Hainhofer (1578–1647) reported that in the Kunstkammer he had seen "a black hill crafted from a touchstone, on top sits an ape with a music book before it. It beats a rhythm and rolls its eyes. Around it sit several animals, all of which are enameled and gold. It looks like the wolf preaching to the geese."[32] Hainhofer was reminded by the object of the popular tale of Reynard the fox, who—according to the numerous iterations of the tale—duped unwitting poultry into listening to him deliver a sermon in order to capture and devour them.[33] That the automaton recalled the Reynard tale to Hainhofer is no wonder. The object is a remix of numerous sixteenth-century images of the duplicitous fox preaching to ducks, geese, and chickens that mobilized the same composition (fig. 27). Because the composition evokes Reynard imagery, and because the central figure is an ape, one might imagine that the Verkehrte Welt Automaton is simply a homespun allegorical warning against false preachers.[34]

Indeed, an ape is an appropriate figure for a pretender. According to a tradition that reaches back to Aristotle, apes are considered baser versions and unworthy imitators of man.[35] Until Isidore of Seville proved otherwise in his influential summa *Etymologiae*, it was generally accepted that the Latin *simia* (monkey) stemmed from the Latin *similitude*—rather than the Greek *simus*, which means "snubbed nose."[36] Many theologians and compilers of fables and romances believed apes to be such masters of mimicry that the term used to refer to them could not but point to this essential feature. Despite Isidore's efforts, the association between apes and mimicry or pretense held strong and was later elaborated on by other influential authors. "The ape . . . has a natural talent for reproducing the things he has seen by means of ridiculous gestures," wrote the English theologian Alexander Neckam. "Woe, woe! How the ape labors to achieve the semblance of the nobility of man."[37] Apes mimic but they do not mimic well. Neckam casts the ape as an actor who overacts, a player who plays his part with too much gusto and too little nuance. The figure of the ape who performs "ridiculous gestures" and whose performance is vulgar also figures the fool. Chapter 19 of Rabelais's wildly popular *Pantagruel*, titled "How Panurge made an *ape* of an Englishman who argued by signs," illustrates this point rather flamboyantly. Rabelais opens his chapter at the onset of a debate. Thaumaste, a "grandissimo scholar" from England, has traveled to Paris with the sole intention of challenging the giant Pantagruel, a scholar of international renown, to an intellectual duel "by signs alone with no talking." However, through a series of deceptions, and to Thaumaste's dismay, the proud Englishman faces not the great Pantagruel but Panurge, Pantagruel's crafty libertine companion. After Panurge attaches "his ample codpiece" and once the audience has fallen silent, Thaumaste begins his opening argument by raising "both hands high in the air separately and, knotting the tips of the fingers of each hand together so as to form what the people of Chinon call a hen's bottom, he struck the nails of both hands together four times, opened his hands and slapped one palm against the other with a resounding smack. Once again bringing them together as before, he clapped his hands twice, then four times, with open palms. He next put them together, stretched and conjoined, as though devoutly praying to God."[38] Rabelais's farce is predicated on undermining Thaumaste's identification as a scholar by way of his clownish behavior, and Thaumaste's solemnity during his performance of a newly invented and unintelligible language of absurd gestures is crucial to the episode's comedic thrust. Behind this idiocy is Thaumaste's abdication of his own reason—an abdication that is a consequence of his uncontrollable passion to best Pantagruel and then Panurge. This illogical

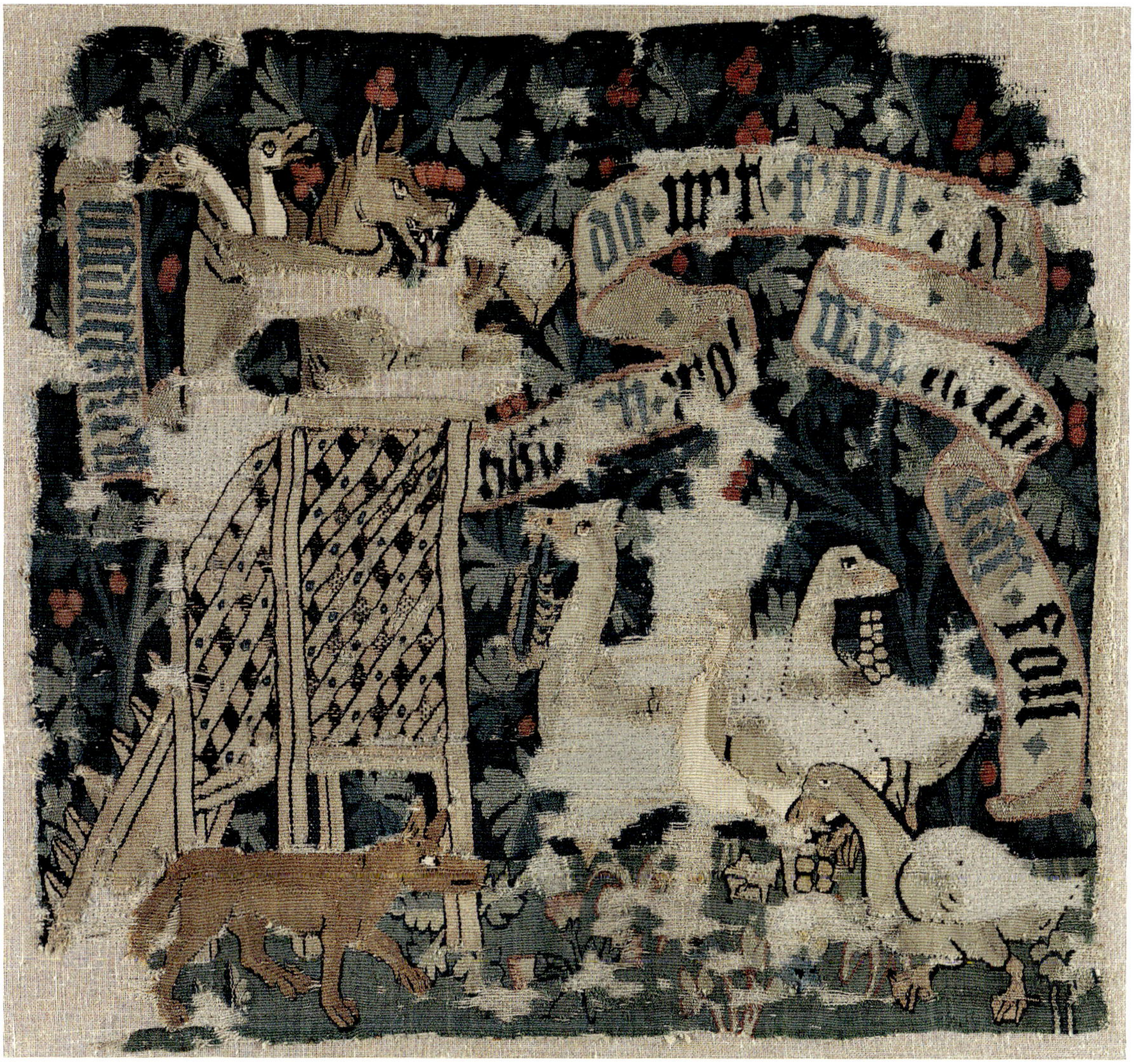

27. *Swiss Tapestry Cushion with Reynard the Fox*, ca. 1525, Burrell Collection, Glasgow.

behavior, and its dramatic presentation before an audience, are transformative.[39] It is how Thaumaste's human status was clouded; it is how he was "made an ape."

The conflation of the ape and the fool was virtually a commonplace during the sixteenth century, especially in the German-speaking world. By the thirteenth century the German term *Affe* (ape) had achieved common currency as a synonym for "fool." Numerous phrases attest to the association between being apish and being a dunce or a loon, someone unbound: *den Affe machen* (to play the fool), *wie ein Affe auf dem Schleifstein sitzen* (to sit like a monkey on a knife-grinder), *wohl vom blauen Affen gebissen* (to be off one's head), *wie vom wilden Affen gebissen da herumtoben* (to jump around like a raving lunatic), *wie einen Affenkäfig* (like a madhouse). In the

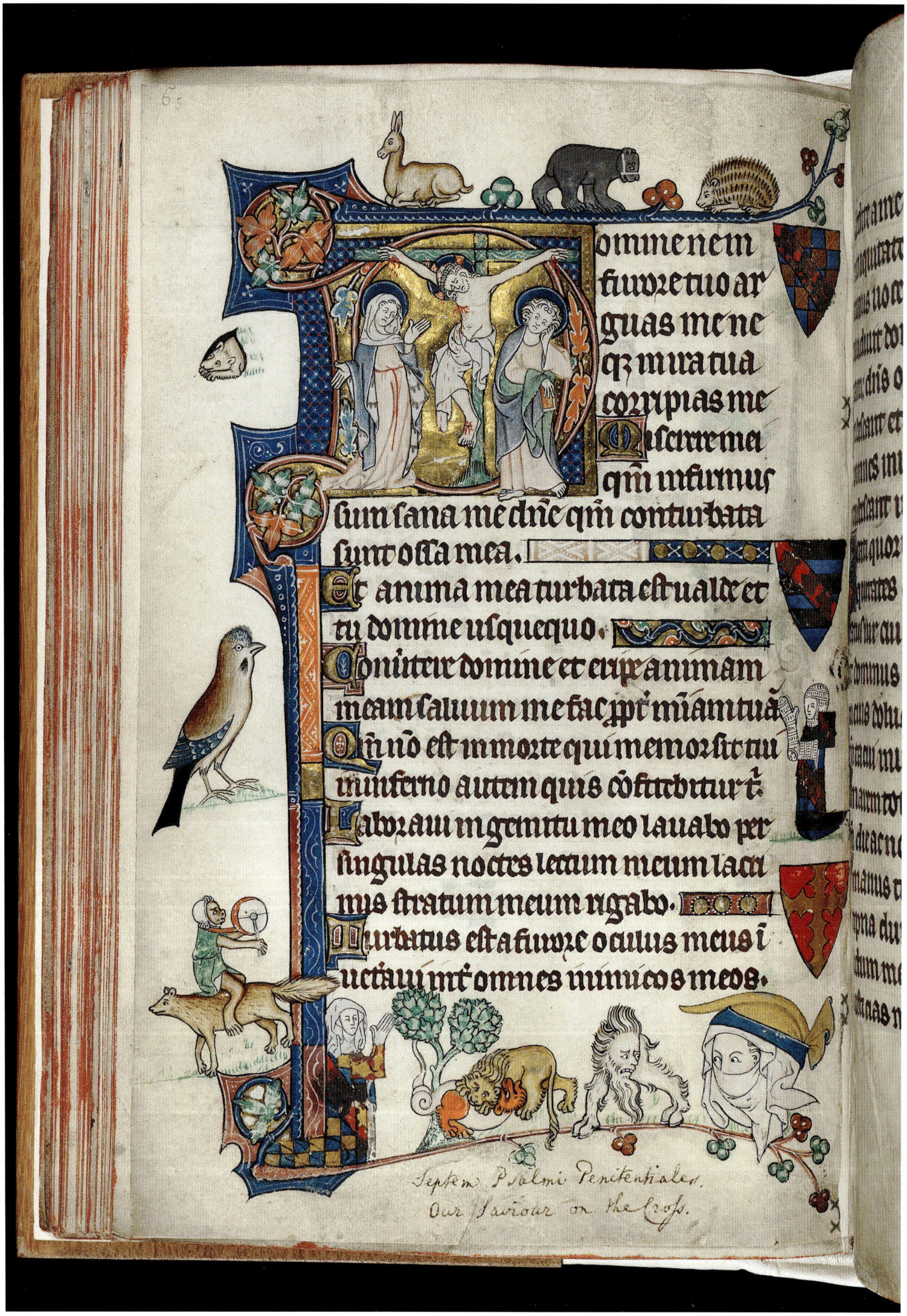

omine ne in
furore tuo ar
guas me ne
q3 in ira tua
corripias me
Miserere mei
qm infirmus
sum sana me dne qm conturbata
sunt ossa mea.
Et anima mea turbata est valde et
tu domine usquequo.
Converte domine et eripe animam
meam salvum me fac ppt miam tuā
Qm nō est in morte qui memor sit tui
in inferno autem quis cōfitebitur t.
Laboravi in gemitu meo lavabo per
singulas noctes lectum meum lacri
mis stratum meum rigabo.
Turbatus est a furore oculus meus i
veteravi inter omnes inimicos meos.
Septem Psalmi Penitentiales.
Our Saviour on the Cross.

medieval period, representations of apes typically insinuate that they are brainless creatures. In the marginalia of a folio from the fourteenth-century Pabenham-Clifford Hours (fig. 28) and a fifteenth-century *Minnekästchen*, or case (fig. 29), from the Bavarian National Museum, apes embody their foolishness by donning a crown, wielding a branch as if it were a scepter (Minnekästchen), and beating a drum while mounted backwards on a dog (Pabenham-Clifford Hours).[40] More in tune with the Verkehrte Welt Automaton is a sermonizing ape depicted on a folio of a fourteenth-century book of hours, now in the Walters Art Gallery (fig. 30). In the register below the text, above a band of foliated and figured roundels, sits an ape in a Gothic throne wielding a crozier, with three other attentive apes clustered around its feet.

Medieval simian imagery is one important source for the subject of the Verkehrte Welt Automaton. The ape's thoughtless, mechanical mimicry surely owes something to its medieval predecessors. While these figurations, which populate the edges of manuscript folios and adorn secular luxury works of art, may seem to predict the golden, enameled, preaching ape and his congregation, they were peripheral to the work of art itself, occurring in places, in the words of Michael Camille, "betwixt and between."[41] By contrast, the Verkehrte Welt Automaton is constructed so as to feature and even animate the ape and its actions. What sort of natural order, not to mention church hierarchy, could sustain an ape preaching? The world represented on this hilltop is doubly topsy-turvy: apes do not preach, and deer do not listen to sermons, whether delivered by apes or men.

28. Pabenham-Clifford Hours (formerly the Grey-Fitzpayn Hours), Ms. 242, fol. 55v, 1315–1320, Fitzwilliam Museum, Cambridge.

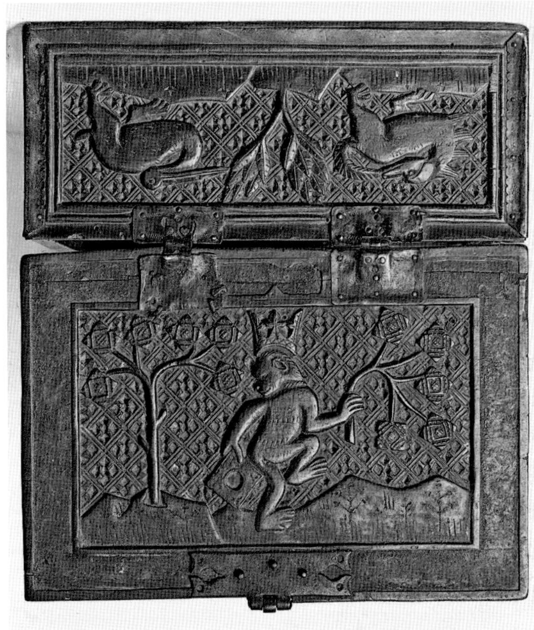

29. Minnekästchen, fifteenth century, Bayerisches Nationalmuseum, Munich.

The Verkehrte Welt

At the time the automaton was made and acquired by the Munich court, German-speaking lands were inundated with verbal and pictorial formulations of the world upside down.[42] The efflorescence of literature, pictures, and objects that depicted a reversal of laws of nature and social hierarchies took its cues from a number of sources—classical mythology, animal fables, exempla, proverbs, folklore, and zoological treatises among them.[43] Broadsheets depicting women beating their husbands, bulls disemboweling butchers, hares hunting men, and monkeys attacking castles were widespread. Tales of ships traversing rocky terrains and peasants drinking spoons were narrated in such widely read texts as Sebastian Brant's *Ship of Fools* (1494) and Hans Sachs's *The Land of Cockaigne* (1530).[44] The world upside down was also featured in religious imagery. Foxes preaching to ducks and farriers shoeing geese appear regularly in the misericords of churches such as Hoogstraten in present-day

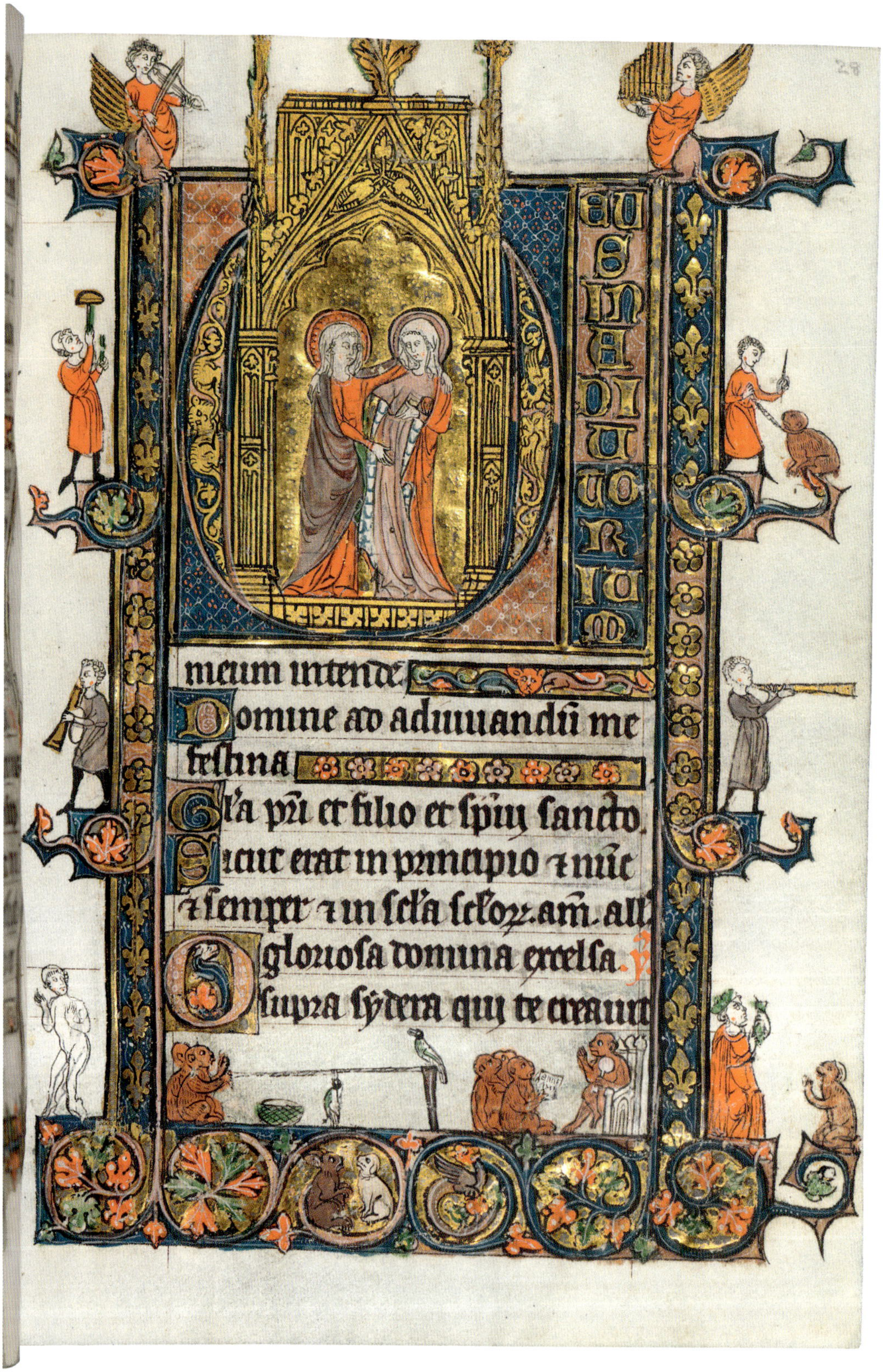

30. Franco-Flemish Book of Hours, Ms. 104, fol. 28, fourteenth century, Walters Art Museum, Baltimore.

Belgium.[45] To a lesser extent, luxury items were also prone to incorporate this imagery. For instance, a silver platter crafted by Abraham Gessner is covered with imagery featuring such reversals, executed with the utmost precision and care (fig. 31).[46]

David Kunzle has proposed that Verkehrte Welt imagery be divided into five types of inversion and one heterogeneous group: human to human; human to animal; animal to animal; animal to element; animal to object; and "animals in incongruous activity." To the first category belong gender and class inversions, such as the common theme of Phyllis riding and subjugating Aristotle. To the second category belong animals that treat humans as animals, such as donkeys riding peasants. To the third category belong images that reverse the roles of animals, for instance the prevalent scene of hares hunting dogs. The fourth category, "animal to element" reversal, is infrequent, but one case can be found in a frame of a Dutch broadsheet in which fish are busily nesting in the tops of trees. The fifth category, the inversion of animal and object, appears frequently in the misericords of churches. One example of this mode of reversal can be found on a misericord in Beverly Minster, a wonderfully foreshortened representation of a cart pulling a horse. A significant portion of Verkehrte Welt imagery falls into the sixth category: animals performing human activities.[47] Dancing bears, cats playing bagpipes, and monkeys riding horses appeared in verbal and visual representations well before the Verkehrte Welt Automaton was crafted.

On one level inversion was intended to unmask a simple social or moral truth—women should obey their husbands, monks should not live lavish lives, nuns ought not be unchaste, gambling is a dangerous ephemeral pleasure, and so on. Yet on another level, although this is less often remarked on in the vast literature on the Verkehrte Welt, these inversions are a mode of imaginative, reflective early modern thought.

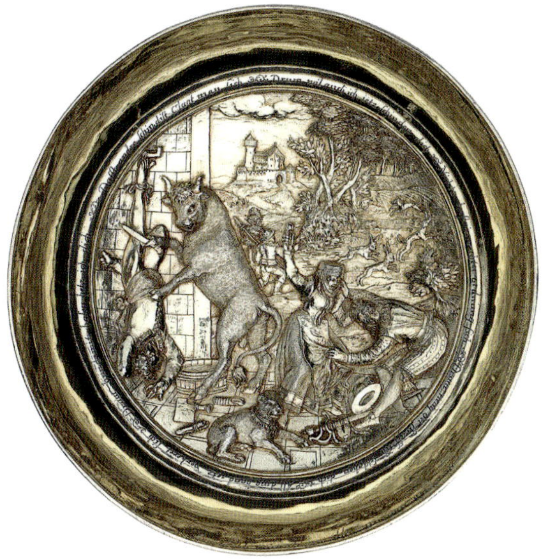

31. Abraham Gessner, Verkehrte Welt Platter, 1580, Schweizerisches Nationalmuseum, Zurich.

They force their audience to consider the opposite of the action or event represented.[48] In replacing the straightforward normative moralistic or social message with its exact contradiction, the Verkehrte Welt employs irony in its classical sense (*quo contraria ostenduntur, ironia est*).[49] This transposition, or *translatio*, of one thing for another is not arbitrary: it "presuppose[s] discovering a *similitude* or something common between the terms, for this is what makes the transfer possible."[50] It is precisely the similitude between the figure of the monkey—an amalgam of mimic, deceiver, lousy performer, and idiot or fool—on the Verkehrte Welt Automaton and the object of its mockery—a preacher—that informs the object's potent message about preaching.

We might conclude that the Wittelsbachs, like other aristocratic patrons and collectors north and south of the Alps, valued the automaton because its fabulous imagery convincingly and compellingly conveyed its maker's *ingenium*, *inventio*, and *imaginatio*.[51] "Fabulous" derives from the Latin *fabula*, synonymous with fiction.[52] But this does not sufficiently explain the significance of the novel, particular, and peculiar aspects of the object's imagery, nor does it shed much light on why the object was acquired by the Munich court and so prominently displayed. To understand its appeal to the Counter-Reformation court in Munich, we must examine another class of images that the automaton readily recalls—representations of Lutheran preachers. Like other Verkehrte Welt images and objects, the Verkehrte Welt Automaton mocks the standing of a social group—Protestant ministers. It goes further than other images and objects of inversion, however, by mocking a newly minted epistemology: the Verkehrte Welt Automaton attempts to undermine one of the ways the reformers claimed to know God.

Figures of Speech

The sixteenth century, during which the German-speaking world produced a density of configurations of the world upside down, was also a time when images of clergymen pulpiteering were manifold and ever present. These images, most of which were produced by Protestant presses and artists, lauded preaching as the foremost strategy to transmit the word of God, as well as a means to connect theological teachings and the concerns of the clergy with the spiritual ideas and practices of their congregations.[53] Scores of woodcuts such as Lucas Cranach the Elder's *Hallowed Be Thy Name* (1527), *The Third Commandment* (1529), and *The Last Supper of Protestants and the Pope's Descent to Hell* (1547); Matthias Gerung's *Evangelical Church Service and Catholic Indulgences* (1546); Georg Pencz's famous *Content of Two Sermons* (1529; fig. 32); and Pancratz Kempf's *The Difference between the True Religion of Christ and the False Idolatrous Teachings of the Antichrist* (1550; fig. 33), to name but a few, became emblems of the evangelical movement.[54] Their imagery, like that of the world upside down, was routine: the laity, young and old, rich and poor, stands or sits in the audience before a pulpit occupied by a gesticulating minister, often with a book laid open before him. Images such as these were reproduced in books, pamphlets, and broadsheets—or "printed catechisms"—that not only publicized the new faith but also conveyed Luther's essential teachings of *sola scriptura*, *sola fide* and *sola gratia* to laypersons.[55]

Of all the printed catechisms mentioned above, Pancratz Kempf's *The Difference between the True Religion of Christ and the False Idolatrous Teachings of the Antichrist* most clearly articulates the Lutheran minister's ability to transfer the word of God to his congregation.[56] In Kempf's complex, two-part image, the rites, wealth, corruption, and idolatry of the Catholic Church are pitted against the cleansed sacraments—the Baptism and Lord's Supper—of Luther's reformed religion. Separated by the column that bisects the image, Luther, pristine in his fur-collared coat, or *Schaube*, is juxtaposed with a hooded monk at the right who has a demon wielding on his shoulder a bellows, a conduit of Satan's message.[57] Wearing a devilish grin, the corpulent clergyman stands in an ornamented pulpit and delivers his sermon (devoted to the teachings of the Antichrist, *Abgottischenlehr des Antichrists*) while pointing toward the monks and church officials below him who carry out what Luther deemed the abuses of the Church.[58] A demon blesses the altar table placed on the naked earth at the center of the composition; a richly adorned priest celebrates a private mass; a bishop consecrates a bell; two monks lay a cowl on a dying man to ensure that he expires as a member of a privileged monastic order. Holding aloft the standard of a local saint, a monk leads a procession

32. Georg Pencz, *The Content of Two Sermons*, 1529, Kupferstichkabinett, Berlin.

33. Pancratz Kempf, *The Difference between the True Religion of Christ and the False Idolatrous Teachings of the Antichrist*, 1550.

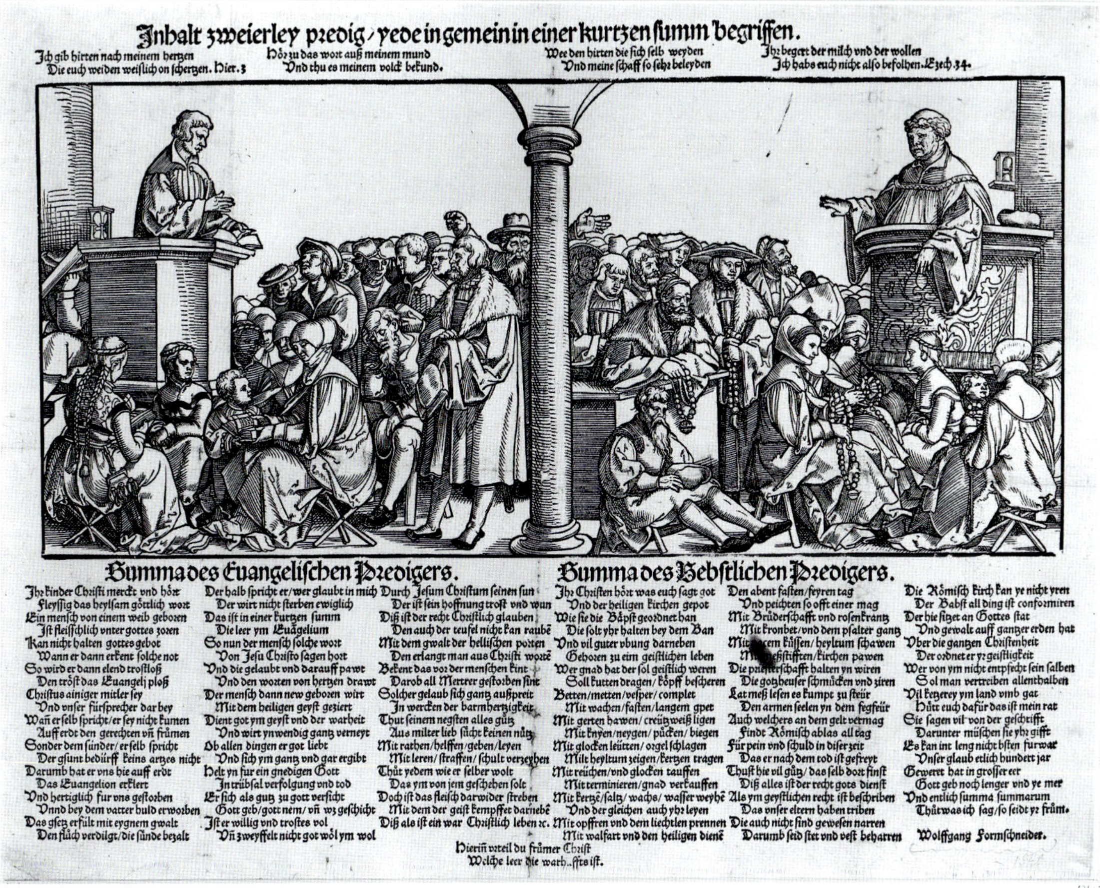
Inhalt zweierley predig/ yede in gemein in einer kurtzen summ begriffen.
Ich gib hirten nach meinem hertzen
Die euch weiden weislich on schertzen. Hier. 3
Hör zu das wort auß meinem mund
Vnd thu es meinem volck bekund.
Wee den hirten die sich selb weyden
Vnd meine schaff so sehr beleyden
Ihr begert der milch vnd der wollen
Ich habs euch nicht also befolhen. Ezech. 34.
Summa des Euangelischen Predigers.
Ihr kinder Christi merckt vnd hört
Fleyssig das heylsam göttlich wort
Ein mensch von einem weib geboren
Ist fleisschlich vnter gottes zoren
Kan nicht halten gottes gebot
Wann er dann erkent solche not
So wird er dann elend trostloß
Den tröst das Euangelj ploß
Christus ainiger mitler sey
Vnd vnser fürsprecher dar bey
Wañ er selb spricht/ er sey nicht kumen
Auff erdt den gerechten vñ frümen
Sonder dem sünder/ er selb spricht
Der gsunt bedürff keins artzes nicht
Darumb hat er vns hie auff erdt
Das Euangelion erklert
Vnd hertiglich fur vns gestorben
Vnnd bey dem vatter huld erworben
Das gsetz erfült mit eygnem gwalt
Den flůch verdilgt/ die sünde bezalt
Der halb spricht er/ wer glaubt in mich
Der wirt nicht sterben ewiglich
Das ist in einer kurtzen summ
Die leer ym Euāgelium
So nun der mensch solche wort
Von Jesu Christo sagen hort
Vnd die gelaubt vnd darauff pawt
Vnd den worten von hertzen drawt
Der mensch dann new geboren wirt
Mit dem heiligen geyst geziert
Dient got ym geyst vnd der warheit
Vnd wirt ynwendig gantz verneyt
Ob allen dingen er got liebt
Vnd sich ym gantz vnd gar ergibt
Helt yn fur ein gnedigen Gott
In trübsal verfolgung vnd tod
Er sich als gutz zu gott versicht
Gott geb/ gott nem/ vñ wz geschicht
Ist er willig vnd trostes vol
Vñ zweyffelt nicht got wöl ym wol
Durch Jesum Christum seinen sun
Der ist sein hoffnung trost vnd wun
Diß ist der recht Christlich glauben
Den auch der teufel nicht kan raubē
Mit dem gwalt der hellischen porten
Den erlangt man aus Christi wortē
Bekent das vor der menschen kint
Darob all Mertrer gestorben sint
Solcher gelaub sich gantz außpreit
In wercken der barmhertzigkeit
Thut seinem negsten alles gůtz
Aus milter lieb sůcht keinen nůtz
Mit rathen/ helffen/ geben/ leyen
Mit leren/ straffen/ schult verzeyhen
Thůt yedem wie er selber wolt
Das ym von jem geschehen solt
Doch ist das fleisch darwider streben
Mit dem der geist kempffet darnebē
Diß als ist ein war Christlich leben zc.
Hierin vrteil du frůmer Christ
Welche leer die warh..ffts ist.
Summa des Bebstlichen Predigers.
Ihr Christen hört was euch sagt got
Vnd der heiligen kirchen gepot
Wie sie die Bäpst geordnet han
Die solt yhr halten bey dem Ban
Vnd vil guter vbung darneben
Geboren zu eim geistlichen leben
Wer gnad hat der sol geistlich weren
Soll kutten dragen/ köpff bescheren
Betten/ metten/ vesper/ complet
Mit wachen/ fasten/ langem gpet
Mit gerten hawen/ creütz weiß ligen
Mit knyen/ neygen/ pücken/ biegen
Mit glocken leütten/ orgel schlagen
Milt heyltum zeigen/ kertzen tragen
Mit reüchen/ vnd glocken tauffen
Mit terminieren/ gnad verkauffen
Mit kertzē/ saltz/ wachs/ wasser weyhē
Vnd der gleichen auch yhr leyen
Mit opffren vnd dem liechtlen prennen
Mit walfart vnd den heiligen dienē
Den abent fasten/ feyren tag
Vnd peichten so offt einer mag
Mit Brůderschafft vnd rosenkrantz
Mit kronbet/ vnd dem psalter gantz
Mit [illegible]cem küssen/ heyltum schawen
M[illegible] meßstifften/ kirchen pawen
Die pri[illegible]sterschafft halten yn wiren
Die gotzheuser schmücken vnd ziren
Lat meß lesen es kumpt zu steür
Den armen seelen yn dem fegfeür
Auch welchers an dem gelt vermag
Findt Römisch ablas all tag
Für pein vnd schuld in diser zeit
Das er nach dem tod ist gefreyt
Thust hie vil gůtz/ das selb dort finst
Diß alles ist der recht gots dienst
Als ym geystlichen recht ist beschriben
Das vnser eltern haben triben
Die auch nicht sind gewesen narren
Darumb seid stet vnd vest beharren
Die Römisch kirch kan ye nicht yren
Der Babst all ding ist conformiren
Der hie sitzet an Gottes stat
Vnd gewalt auff gantzer erden hat
Vber die gantzen Christenheit
Der ordnet er yr geistligkeit
Wer von ym nicht entpfecht sein salben
Sol man vertreiben allenthalben
Vil ketzerey ym land vmb gat
Hůtt euch dafür das ist mein rat
Sie sagen vil von der geschrifft
Darunter müschen sie yhr gifft
Es kan int leng nicht bsten furwar
Vnser glaub etlich hundert jar
Gewert hat in grosser eer
Gott geb noch lenger vnd ye mer
Vnd entlich summa summarum
Thůt was ich sag/ so seidt yr frům.
Wolffgang Formschneider.

631 - 7

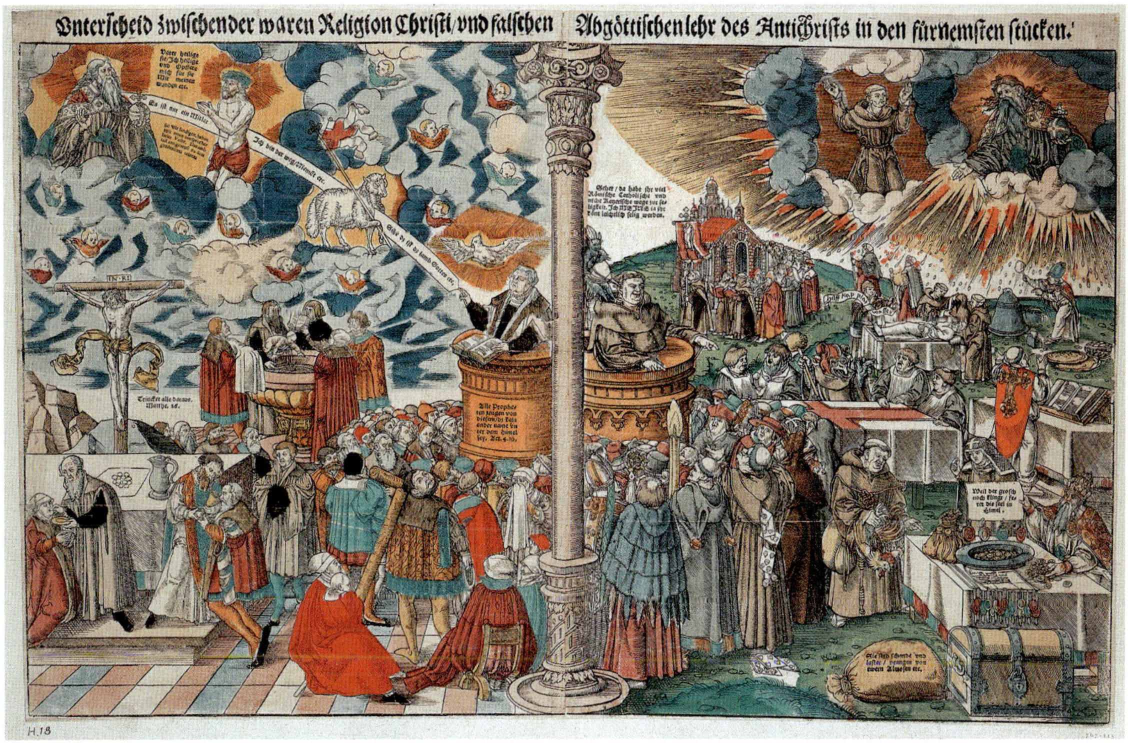
Vnterscheid zwischen der waren Religion Christi/ vnd falschen Abgöttischen lehr des Antichrists in den fürnemsten stücken.

H.18

of pilgrims around the church in the background. In the right foreground, wearing the three-tiered papal tiara, Christ's vicar on earth is selling indulgences. He fondles a pile of coins with his left hand, while in his right he holds aloft an indulgence paraphrasing the infamous lines of Johann Tetzel from 1517: "Sobald der Gülden im Becken klingt / im huy die Seel im Himmel Springt" ("When the coin in the coffin rings / the soul from purgatory springs"). Encircled by clouds and kneeling beside God the Father, Saint Francis, plainly displaying his stigmata to the viewer, offers himself as intercessor, prompting the divine wrath of fire and brimstone.

In the other, left, half of the image, Luther is identified as a prophet by the text inscribed on the pulpit, "Alle Propheten zeugen von diesem dz sein ander name unter dem Himel sey" ("All prophets testify to this one, that his other name is in the heavens"). Luther solemnly places his left hand on the book of Scripture, while his right hand extends to make contact with an outstretched banderole containing a procession of words that concludes in the body and orb of God the Father in heaven. Reading from above, the first of these texts states, "Es ist nur ein Mitler" ("There is only one mediator"). The text between Christ and the Paschal Lamb reads: "Ich bin der weg Niemant etc.," an abbreviation of John 14:16 ("I am the way, the truth, the life: no man cometh unto the father but by me"). The third text starts anew with John 19:29, "Gibt es ist das lamb Gottes etc." ("Behold, this is the Lamb of God, who takes away the sin of the World"). A passage of text to the left of Christ's head (the suggestion is that the words emerge from his mouth) reads, "Vater heilige sie / Ich heilige und opffere mich für sie. Mir meinen wunden etc." ("Holy Father save them, I sacrificed myself for them with my wounds"). Reiterating this salvific plea, the text below reminds the viewer, "So, wir sundigen haben wir einen wortsprecher beim Vater. Darumb last uns gewost zu dem gnadenivol treten" ("If we sin we have an advocate before God, so let us turn in consolation to this means of grace"). Directly beneath this message, communion is given to a man and a woman kneeling at the altar table, behind which the crucified Christ rises above the landscape. To his right we read Matthew 26:27, "Trinkt alle daraus" ("Drink from it, all of you"). The passage from Matthew's Gospel is strategically placed to bridge the scene of the Eucharist to the second sacrament of Luther's church—baptism. By picturing the channels through which preachers receive and deliver God's invisible message and by juxtaposing this invisibility with the wickedly visible and material mechanisms of salvation institutionalized by the Latin church, Kempf's woodcut asserts that Luther, and, by extension reform preachers in general, are the shepherds of Christ's kingdom, because "Faith cometh by hearing, and hearing by the word of God" (Romans 10:17). It is not surprising, then, that the image tickles one's ears and that it does so by including a modest and condensed text. Kempf's image not only shows how God's word is transmitted: it forces the literate viewer to hear the inaudible word of God, but to hear it in his or her own voice. "God puts words in our mouth as if he himself were singing in us," wrote Calvin in the introduction to his Psalms.[59]

Joseph Koerner's study of the *Wittenberg Altarpiece* (1547) by Lucas Cranach the Elder (1472–1553) adduces the theological and political underpinnings of images of Protestant ministry. Koerner discusses the wide dissemination of images of preaching, which he understands as attempts to picture an ideal communicative relationship between the Protestant minister and his congregation (*Gemeinde*) of believers. Standing in stark contrast to the monstrances, papal seals, coins, rosaries, and reliquaries that populate scenes of Catholic materialism and ritual, printed images of preaching disavow the power of visual persuasion integral to Catholic practices. No longer dependent on the *heilbringende Schau* (salvific display) and the *Schaufrömmigkeit* (visual piety) of the Latin church, the reform church relies only on words—the Scripture and sermons—to convey

Christ. The depiction of these relationships rests on the assumption that a given message can be transmitted immediately and clearly to the heart and conscience of each individual believer.[60] Inwardness is publicly performed, paradoxically.

Viewed in light of the early sixteenth-century pictorial conventions for representing the act of preaching in the context of the Reformation, and of Koerner's work, the Verkehrte Welt Automaton offers a parodic commentary on reform preaching (and its representation) by way of silently enacting it. It appears to mock the premise that meaningful communication can occur between minister and congregation. Like the majority of moral comedy, the object sends an ironic, clear, and well-defined message: the reformed minister's ability to transfer the word of God to the populace is comparable to the ape's claim to be able to communicate with deer. This hilltop farce therefore stages not merely a reinterpretation or simple rephrasing of the printed catechisms but an entirely different model of oral transmission that places the Protestant sermon outside the realm of religiosity. On the one hand, the automaton parodies and thereby condemns a social phenomenon and, like other Verkehrte Welt imagery, relies on absurdity as a comedic resource. On the other hand, by twisting the mechanics of Protestant ministry into mindless repetition and thoughtless gestures, and by representing the audience as an unlikely assemblage of uninterested and discrete listening bodies, the automaton denies the notion of verbal efficacy—one of the most crucial components of Luther's Reformation, if not the single most crucial.

The Kingdom of Listeners

"The office of preaching is second to none in Christendom."[61] This statement on the glorification of preaching was written by Martin Luther and intended for his freshly formed congregation of Protestants in Wittenberg. Writing in 1522, the young Luther modeled his sermon *On the Office of Preaching and of Preachers and Hearers* after John 10:1–11, in which Christ recounts the parable of the shepherd.

> Verily, verily I say unto you, he that entereth not by the door into the sheepfold, but climbeth up some other way, is the same as a thief and robber. But he that entereth in by the door is the shepherd of the sheep. To him the porter openeth; and the sheep hear his voice; and he calleth the sheep by name and leadeth them out. And when he putteth forth his own sheep, he goeth before them, and the sheep follow him: *for they know his voice.* And a stranger they will not follow, but will flee from him for they know not the voice of strangers (my emphasis).

Like the Verkehrte Welt Automaton, Luther's sermon could be construed as a warning against false preachers. Luther knew well that for the Church of Rome, the pulpit served exclusively to lay down rules of good behavior and proper spiritual actions, since the seven sacraments were considered the primary mechanisms of salvation. Luther's foundational sermon on the office of ministry was not inspired by a conviction that the church's clergymen were pretenders, however, but by his belief in the efficacy of the spoken word. Later in the sermon he writes: "Let it be called 'coming' when one preaches right; the approach is spiritual, and through the word—upon the ears of listeners [*Höreren*], the preacher comes at last into the sheepfold—the heart of believers. Christ says that the shepherd must enter the door; that is, preach nothing but Christ, for Christ is the door onto the Sheepfold."[62] Luther reiterated this claim one year later, in the preface to the *Leisnig Order* (1523): "Every householder and his wife shall be duty-bound to cause the wholesome and consoling Word of God to be preached to them, their children, and their domestic servants, so that the Gospel may be impressed [*eingebildet*] on them for their betterment."[63] He further refined this notion at the end of his career, in a sermon on the Eighth Psalm in 1545: "Christ's kingdom is a hearing kingdom, not a seeing kingdom; for the eyes do not lead and guide us to where we know and find Christ, but rather ears do this."[64] In emphasizing the exchange between the preacher (speaker) and the congregation (hearers) Luther proposes that oral communication—"who says what to whom in what channel with what effect"—is a salvific mechanism.[65] Luther's advocacy of the pulpit did not fall on deaf ears. In the words of Robert Scribner, "for Protestants, 'hearing the Word' became virtually a third Sacrament alongside Baptism and the Lord's Supper."[66]

The notion that preaching was a means of mass communication to reach a religious base is not Luther's own: it was already mobilized in the late fourteenth, fifteenth, and early sixteenth centuries by charismatic figures such as Jan Hus (1369–1415) and Geiler von Kaysersberg (1445–1510).[67] In addition to eradicating the mystical order of ecclesiastical hierarchy and asserting a concord between the infinite and language, what is novel about Luther's approach, and what makes it so pivotal in the rise of the modern evangelical movement, is that it made salvation largely dependent upon speech acts. For Luther, words—nouns, verbs, adverbs, and adjectives—convey pure grace and enable its passage from Christ to his flock. Luther thus endows preachers with spiritual authority and listeners with the ability to comprehend and retain Scripture's message, which in turn makes them the direct recipients of God's Word, thereby saving them.

In formulating this radical salvific mechanism, Luther had to contend with eight hundred years of eloquent church writing concerning the inability of language to articulate the sacred mysteries of God and of the established church. For theologians language was not only insufficient for capturing the ineffability of the mysteries of the Church; it was also vulnerable to distortion, opacity, and faulty transmission.[68] According to the Church in Rome, sight, touch, and taste were the layperson's primary tools to understand immanence and achieve salvation.[69] Consuming the host, witnessing the weeping cult statue, handling a relic, praying before an altarpiece of Christ, the Virgin, or a saint, beholding the monumental sacred architecture: these experiences proclaimed the *Praesentia* of the Catholic Church, stood as evidence of the efficacy of its rites, and validated the sacred economy on which the institution relied. Language and all of its explanatory, exorcistic, and transubstantiative powers were restricted to the ordained. For instance, translating the Canon into the vernacular was prohibited.[70] Protestant authors often discussed the Catholic Church's silence in their sermons on salvific matters. In a sermon on Luther's life published in 1566 by Johann Mathesius, he reports that before he was introduced to Luther's teachings he was ignorant of the significance of the Eucharist: "I know for certain that, before I came to Wittenberg, I never in my lifetime learned of the forgiveness and comfort that one gets by consuming the body and blood of Christ in faith; neither in church or in school was a word heard recollecting this."[71]

To be sure, preaching was practiced by members of the Franciscans, Carmelites, and other orders in the late Middle Ages, and their sermons addressed a wide range of topics—theological dilemmas, social issues, vice, virtue, punishment, glory, and good works. Yet despite the prevalence of preaching in certain orders there was consensus neither on the function, form, or purpose of ministry nor on its place within the liturgy. Some days a sermon was delivered before the singing of the Gospel, other days after the singing of the Creed. "It is no exaggeration," writes John O'Malley, "to affirm that one aspect of the general religious onus of the period was a crisis in preaching."[72]

No order was more convinced of preaching's significance than the Jesuits. Preaching was listed as the first ministry in both versions of the Jesuit

Formula, and in the ranking system laid out in the *Constitutions*, preaching was placed above giving the *Spiritual Exercises* and hearing confession. It was, Jesuits believed, an apostolic mandate. "And he commanded us to preach unto the people, and to testify that it is he which was ordained of God to be the judge of quick and dead" (Acts 10:42). Like the apostles on whom they modeled themselves, the Jesuits preached widely and openly in churches before and after Mass, and in their imitation of Christ they preached *en plein air*, on a hill or in the streets. It was not uncommon for a Jesuit to stand behind the pulpit three or four times a day. Like their Catholic colleagues, however, they left no official message regarding what their preaching should center on, nor did they contend, like Luther, that the act of preaching led to the salvation of their listeners.[73]

Babel

Counter-Reformation propaganda signals the Church's distrust both of the laity's ability to comprehend the proceedings and of preaching in general. One fascinating response to Luther's elevation of the office of preaching and the spoken word belongs to the early phase of the Counter-Reformation. On the title page of *Most Worthy Sacrament of the Altar* (1529) by the priest and humanist Johannes Cochlaeus (1479–1552), Hans Brosamer (ca. 1490–1554), a precocious student of Lucas Cranach the Elder, casts Luther as the seven-headed hydra from the Book of Revelation (fig. 34).[74] What is most crucial about this image, however, is not its rendering of Luther as an apocalyptic figure, but rather the claims it makes about his ability to convey Scripture's message. Each of the seven heads portrays Luther as a different persona—doctor, saint, infidel, priest, zealot, bureaucrat, and Barabbas. By throwing the stability of Luther's identity into question and by equating these different guises with the head of the "red dragon" of John's revelation, the title page attempts to undermine Luther's spir-

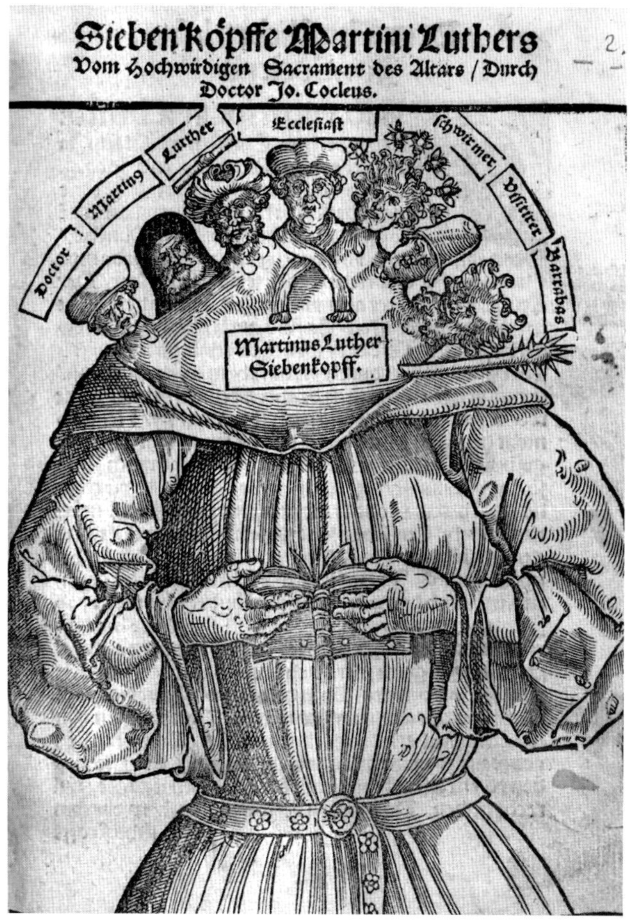

34. Hans Brosamer (attributed), *The Seven Heads of Martin Luther*, 1529, title page from Johannes Cochlaeus, *Sieben Kopfe Martin Luthers: Vom hochwirdigen Sacrament des Altars* (Leipzig: Valentin Schumann, 1529).

itual, theological, and political authority. Yet insofar as this print endeavors to demonstrate Luther's hypocrisy, it also makes a broader claim about the act of preaching. Poised frontally, Luther grasps the Gospel, from which all of his various heads preach. The cacophony transforms the red dragon of Revelation into a "monster of confusion."[75] The composition implies that Luther is directly addressing the viewer, rendering her the audience of his Babel. Like the Verkehrte Welt Automaton, the print represents the breakdown of communication—paradoxically, by way of a sermonizing figure.

Seen in light of the Counter-Reformation court, the confessional and political atmosphere in which the Verkehrte Welt Automaton was made, and the nature and imagery of widespread images of pulpiteering, the automaton no longer seems *quite* so strange. What is most remarkable about the Verkehrte Welt Automaton is the manner in which it invokes and rephrases this preaching imagery. The Protestant minister becomes an animated golden ape; the congregation is transformed into resplendent, indifferent deer; and the reformed church is supported by a hilly mound of trees. By these transpositions, the automaton suggests that the event it enacts is itself preposterous. By means of an internal mechanism that animates the object, it attempts to engender speech itself; it vilifies repetition, demonstrates the routinization of charisma, and signals the futility of attempting to convey Christ through words. To put it another way, the office of preaching, which appears in printed catechisms as trustworthy, effective, and sensible, is unmasked by the Verkehrte Welt Automaton as being unreliable, pointless, and inane. The object's bewildering imagery, setting, and silence are amplified by its addressing the beholder as if she were party to the incomprehensible sermon. The viewer becomes a passive auditor to the silent, repetitive preaching in a bewildering landscape, just as the Catholic Church and, by extension, the Wittelsbach dukes, construed the simple folk. As a representational machine, the automaton rejected the primacy of the sermon in conveying Christ. Unlike Brosamer's *The Seven Heads of Martin Luther*, the Verkehrte Welt Automaton does not make this claim unambiguously. Whereas Brosamer, like the majority of Reformation and Counter-Reformation propaganda authors, relied on the printed word to spell out his message, the Verkehrte Welt Automaton relied on its imagery, composition, and most of all its animation.[76] Remarkably, it becomes a phenomenon of communication—a silent object that endeavors to condemn words.

Five

Habsburg-Ottoman Diplomatic Machinery

From 1547 to 1593 Habsburg rulers of the Holy Roman Empire were forced to pay the sultan of the Ottoman Empire an annual tribute. A result of a truce signed by Archduke Ferdinand I (1503–1564) and Sultan Süleyman the Magnificent (1494–1566) on July 19, 1547, this transcultural transfer of wealth and goods was referred to by German speakers as the *Türkenverehrung*, or "Turkish gifts."[1] A diplomatic euphemism, the expression implied that the Habsburg rulers—Archduke Ferdinand I (later Holy Roman Emperor Ferdinand I, r. 1558–64), Holy Roman Emperor Maximilian II (1527–1576, r. 1564–76), and Holy Roman Emperor Rudolf II—willingly gave unreciprocated gifts to the Ottomans: the rhetoric attempted to transform the obligatory nature of tribute into an act of disinterested prestation.[2] Each year the tribute payment of forty thousand ducats was supplemented by precious objects—silver and gold vessels, clocks, and clockwork automata.[3] Although a great number were sent to high-ranking Ottoman officials over the forty-six-year period, very few automata intended as addenda to the payments survive.[4] This chapter is concerned with three examples (figs. 35, 36, and 37). All were manufactured by clockmakers in the Free Imperial City of Augsburg sometime between 1565 and 1580. All are of average size—for a sixteenth-century automaton—measuring about forty centimeters in height and twenty centimeters in width. All have a standard program, lasting anywhere from 60 to 120 seconds—depending on how tightly they were wound. And remarkably, all represent high-ranking Turkish officials in ceremonial processions.[5]

This chapter addresses these automata as instruments of the Holy Roman Empire's foreign policy at a time when it was intensely competing with the Ottoman Empire for political and religious hegemony over the former Roman Empire.[6] It considers the imagery and animation of the automata in relation to the diplomatic relationship forged between the Holy Roman Empire and the Ottoman Empire in the second half of the sixteenth century.[7] The question of how automata were supposed to function in the diplomatic theater becomes central, providing an opportunity to situate European court culture and technology in relation to European and Ottoman imperial expansion.[8] The aim is to use these automata to reveal a dimension of complexity in Ottoman and Habsburg diplomatic relations that has heretofore not been addressed, as well as to show an instance in which the imagery and movement of automata relayed explicit political statements outside the context of a German-speaking court.

Ottomans, Automata, and Temporal Integration

At the outset of the Türkenverehrung, in 1548, Archduke Ferdinand I made the imperial

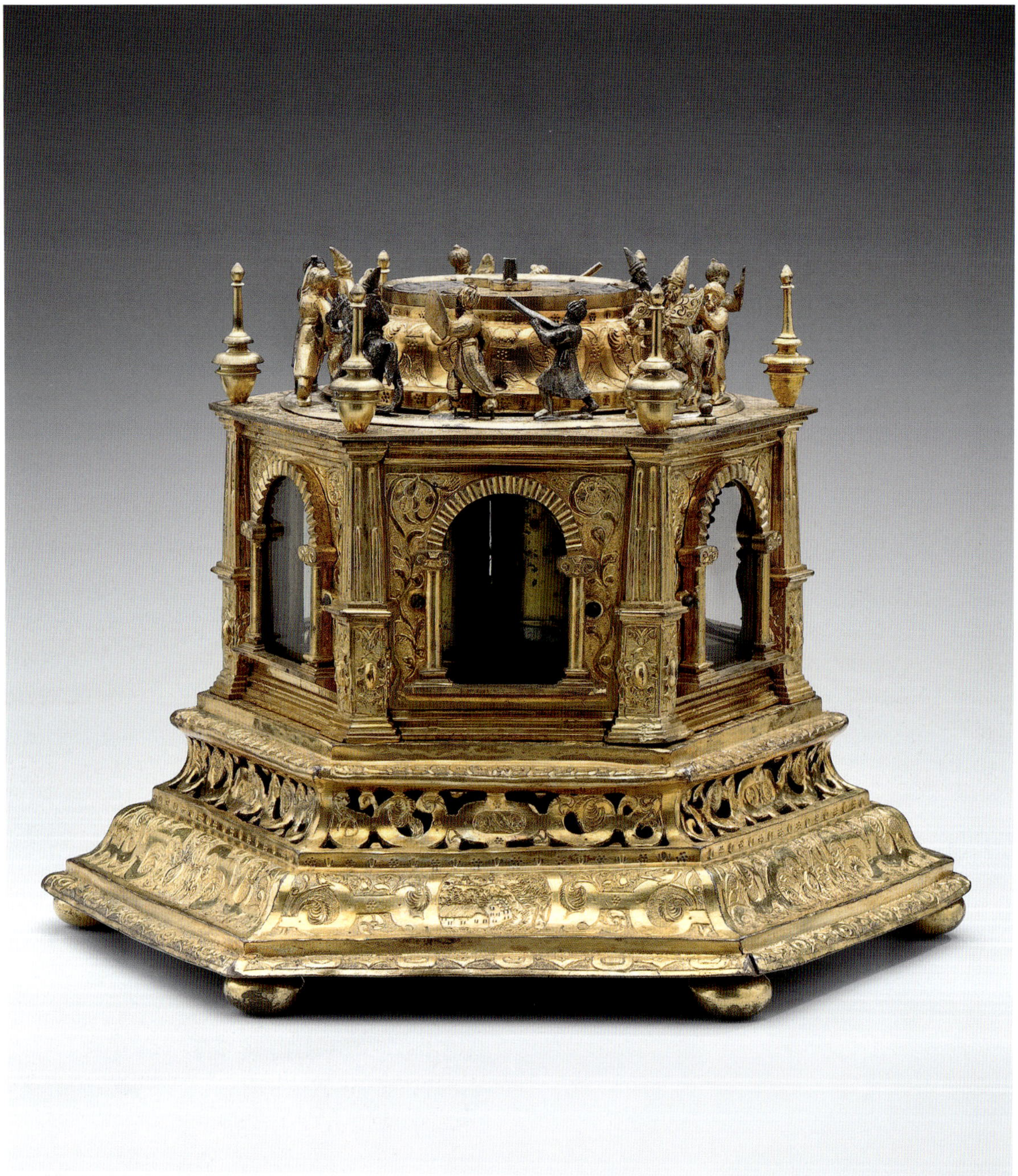

35. *Pashas on Horseback*, 1580–1590, Newark Museum of Art, Newark, New Jersey.

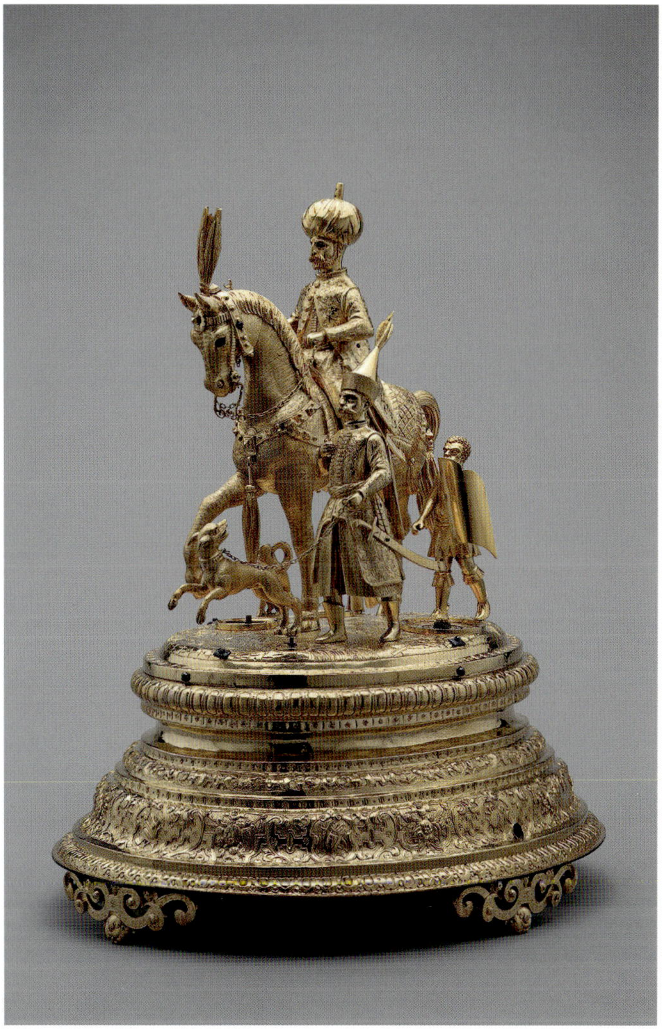

36. *Sultan on Horseback with Attendants*, 1590, Kunsthistorisches Museum, Vienna.

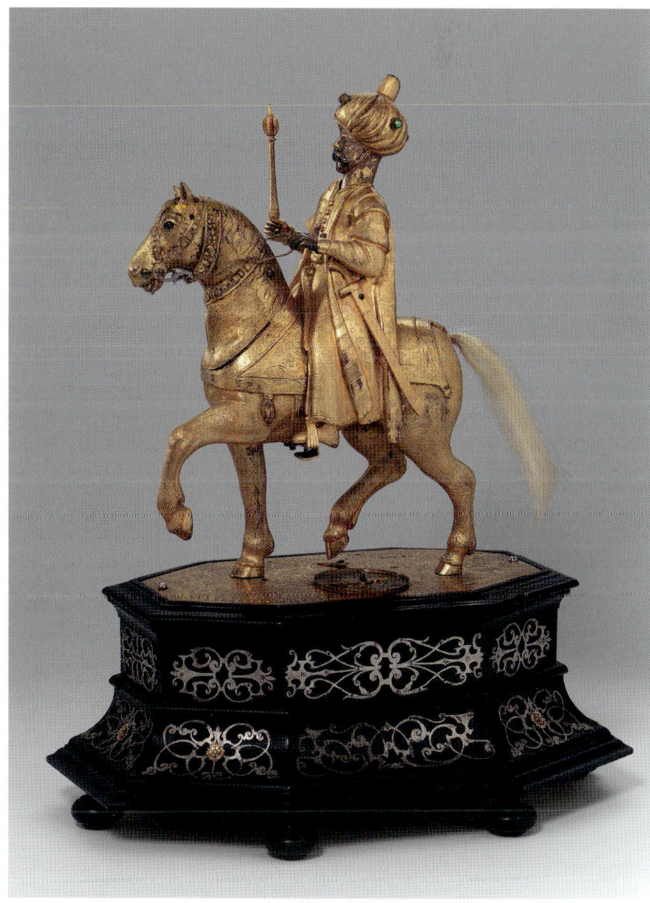

37. *Sultan on Horseback*, 1580–1595, Mathematisch-Physikalischer Salon, Dresden.

governor of Swabia responsible for overseeing the Turkish gifts, and this post remained in the care of that office until 1593, when the annual tribute was halted.[9] The governor exchanged and raised money for payment to the Turks.[10] He was also entrusted with commissioning clockmakers to create the automata that would supplement the funds. The governors must have shown clockmakers prints or drawings of Ottoman ceremonies, for several of the automata betray a reliance on such models. Jan Swart van Groningen's prints of Ottomans on horseback come immediately to mind, as do the prints of the Ottoman court produced by the Danish printmaker Melchior Lorichs. Yet we have little evidence to suggest that Lorich's prints were disseminated before 1620, even though he was in Istanbul from 1556 to 1582. Otto Mayr and Klaus Maurice have suggested that the sultans on the automata in Vienna and Dresden are likely based on *The Procession of Süleyman the Magnificent through the Hippodrome* (1533; figs. 38 and 39), a popular and widely circulated woodcut by Pieter Coecke van Aelst, the younger (1502–1550).[11] It is also possible that Ferdinand, Maximilian, or Rudolf provided models for the automata conveying Ottoman processions. The inventory of the Archducal Kunstkammer at Ambras (1596) records "a long scroll on which the Turkish emperor rides to the Churches," and the inventory of the imperial Kunstkammer in Prague (1607–11) contained "a hand painted book with all types of Turkish costumes and ceremonies."[12]

The imperial diplomat Stephan Gerlach (1546–1612) tells us that in 1590 Holy Roman Emperor Rudolf II gave Sultan Murad III an automaton in the shape of a castle. When the object was activated, the gates of the structure opened and a sultan emerged on horseback, followed by several pashas, also on horseback. The figures, which were made entirely of silver, rode in a circle before disappearing behind a second gate embedded in the face of the fortress.[13] Although much simpler in its composition, a surviving automaton, now in Newark, approximates Gerlach's description (fig. 35). The octagonal silver base of the object is festooned with eagles in flight and aniconic curved motifs, which recall the Ottoman crescent. Resting atop the base is a second octagonal structure, with engaged pillars on all corners. Each pair of pillars frames an arch, through which one can see the object's movements. Swirling vegetal motifs crawl up the sides of the arches and terminate in a tulip in full bloom. All of this extensively wrought and ornamented silver was the foundation for the animation of Ottoman ceremony and pomp. Four figures proceed on horseback in pointed caps, possibly *külahs*, carrying swords on their hips. These figures are the cardinal points of the procession, which moves clockwise in a circle around the raised timepiece. In front of and behind each equestrian figure strut two turbaned figures clutching spears. Although neatly organized around the two types of figures, the automaton is dominated by the riders, whose social status is greater than those of their earthbound counterparts. As well as conveying the look, action, and social rank of an Ottoman official, these features promoted the object's purported novelty, because the recipient was most likely intended to recognize a version of himself in the object.

Two extant automata evidence a similar configuration. An automaton of a sultan on horseback, now in Vienna, depicts the Ottoman official atop a richly saddled steed and accompanied by an attendant holding the leash of a dashing hound, a small sub-Saharan African holding a shield, and a sultan in miniature grasping the chain of a seated ape. When the automaton is activated, the horse taps the sumptuous silver base with its left front hoof and rolls its eyes while the sultan turns his head from side to side, as if looking to the crowd that lined the streets during the processions through Istanbul. By giving the sultan such a varied entourage, the object signals the Ottoman Empire's expansion into the continent of Africa and re-performs, for Ottoman

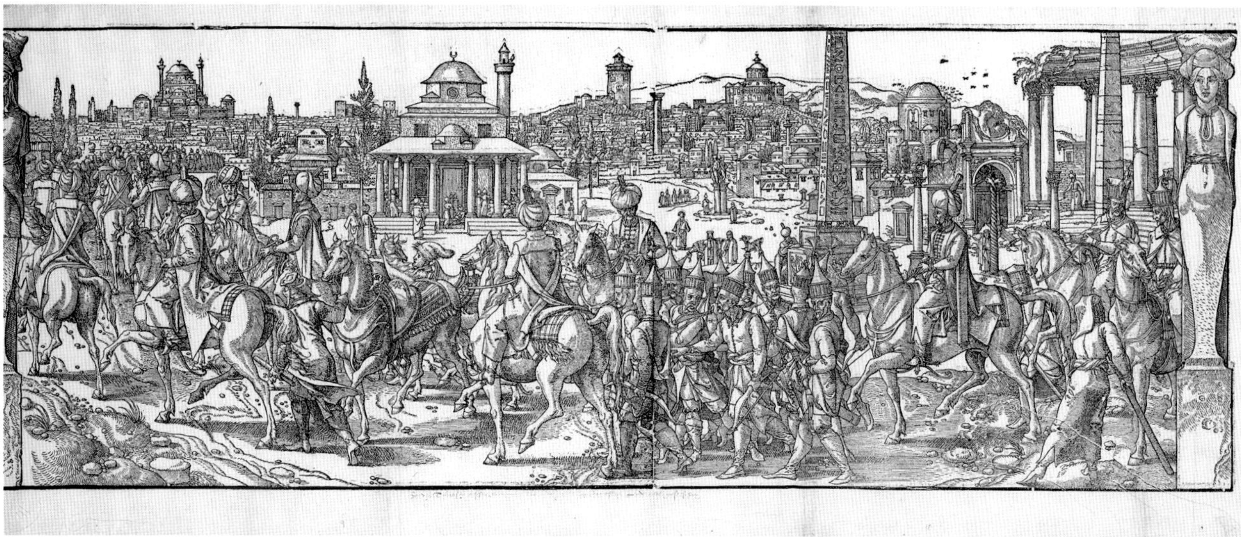

eyes, the ceremonial pomp at events such as royal circumcisions.[14]

Though simpler in its iconography, the automaton housed in Dresden was a multimedia affair. The work is distinguished by the inclusion of horsehair for the mount's tail, and leather or rope reins, now missing, that probably extended from the empty rings at the base of the horse's bit and terminated in the left hand of the sultan. Here the horsehair does not represent a horse's tail in the same way that the gilt body represents the sultan. It differs because it is the thing itself (the horsehair), not a signifier, that refers to an absent referent. This was not the first time props were used to fabricate lifelikeness. Donatello's equestrian portrait of Gattamelata in Padua (1453) included spurs that had been worn by the Venetian condottiere. Figures that formed the extravagant program of Holy Roman Emperor Maximilian I's tomb held real candles and daggers, and the sculptures that populated the reconstruction of the Holy Sepulchre at Varallo wore clothes and dined at tables before dishes of roasted lamb.[15] But unlike Donatello's equestrian portrait, or the "colossal puppets" of Maximilian's tomb, or the statues on the Sacro Monte, the accoutrements attached to the automaton in

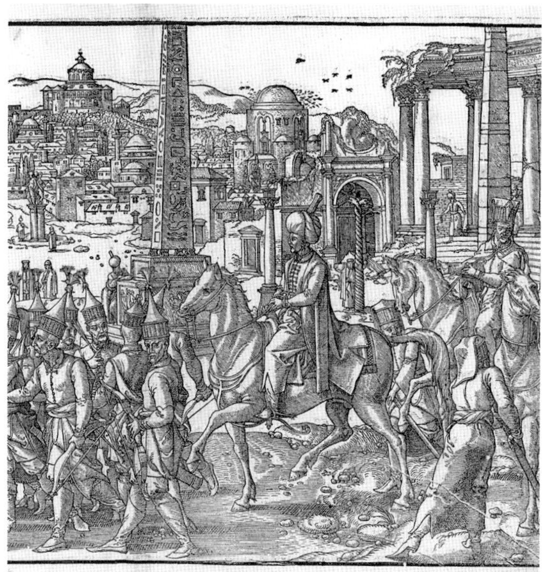

38. (*top*) Pieter Coecke van Aelst, *Procession of Süleyman the Magnificent through the Hippodrome*, 1533, The Metropolitan Museum of Art, New York.

39. (*bottom*) Pieter Coecke van Aelst, *Procession of Süleyman the Magnificent through the Hippodrome* (detail), 1533, The Metropolitan Museum of Art, New York.

Dresden did not merely make the object *appear* as though it were capable of doing something, but rather accentuated what it did. As the horse raised and lowered its head, the reins slackened and tightened while the tail trailed behind as the figure lurched forward.

Processions

Ottoman ceremonial processions were lengthy theatrical displays, rigorously choreographed and highly charged, that made metapolitical claims about the sultan's absolute sovereignty and his relationship to his court, his subjects, and foreign powers.[16] This very public form of statecraft was often the only opportunity for foreign diplomats to catch a glimpse of the "Grand Turk."[17] Every Friday the sultan—astride his imperial steed, with his entourage of viziers, commanders of the imperial guard, foot soldiers, sword and standard bearers, holy men, dwarves, hounds, camels, and falcons—slowly wound his way from the Imperial Gate of the Topkapı Palace, down the Divan Yolu (Council Road), to a series of monumental royal mosques.[18] Additionally, during the celebration of a rite of passage such as the ascension of a new sultan, the wedding of the sultan's sister or daughter, the presentation of a new born royal child, or the circumcision of a royal son, the sultan and close to one thousand Ottoman officials paraded from Topkapı Palace to the Hippodrome, where tents, stages, and seating were erected for public festivities in honor of the event.[19] Often these public festivities were complemented by boat processions along the Golden Horn and the Bosphorus.

Numerous written accounts by European merchants and diplomats praised the elaborateness of the sultan's processions through the city.[20] In printed pictorial accounts of the event, the sultan, atop his horse and surrounded by his entourage, was a kind of ideogram for Ottoman power.[21] Yet in representing these events, artists had to segment into static components the flow of images that processions of this sort generated—panting dogs, bouncing horsehair standards, fluttering, disembodied heron feathers, marching golden dwarves, flapping falcons, and prancing, armored horses. And by doing so the most crucial elements of the procession—movement and the passage of time—were excluded. Artists tried to compensate for this. One strategy was to manufacture images on a large scale, to both include as many participants in the procession as possible and to evoke the passage of time.[22] The woodcut *Süleyman's Procession through the Hippodrome* (1553; fig. 39), by the Flemish printmaker Pieter Coecke van Aelst, the younger, exemplifies this tendency.[23] Spanning almost fifteen feet in length, Coecke van Aelst's frieze—which the artist claimed was based on his own eyewitness account of a procession in 1533—depicts seven scenes and was printed on no fewer than ten separate sheets, each measuring 35 by 87 centimeters.[24] In the sheet that incorporates the sultan, Coecke van Aelst chose to depict one of the most dramatic moments in the procession, if not the most dramatic: Süleyman parading through the Hippodrome.[25] Süleyman, in strict profile atop his horse—a posture echoing monumental imperial Roman equestrian portraits, Byzantine coins, and Renaissance medals—is strategically shown at the moment before he crosses in front of the Pharaonic obelisk of Thutmose III (1479–1425 BCE) that was brought to Constantinople by Emperor Theodosius (347–395) in the fourth century CE. This striking placement allows the artist seamlessly to align the two historical eastern empires of Egypt and Byzantium with the Ottoman Empire. Yet insofar as this compositional strategy appears laboriously staged, the rhetoric of the remainder of the image softens its artfulness by suggesting that it represents a random *punctum temporis*.[26] Coecke van Aelst went to great lengths to render almost all the figures in the sheet as if they had been instantaneously arrested in their own idiosyncratic activity. Compare Süleyman's statuesque posture to that of the figure in the right foreground, turning his back to the viewer as he gracefully lifts and extends his left foot. Or consider the two figures on horseback behind the sultan, who enter from stage left. The direction they face is indicative of their movement through the image, but when the artist placed the caryatid on the edge of the sheet, he obscured the horses' bodies. This partial masking enhances the horses' action. The picture relies not only on a lack of geometrical clarity to suggest instantaneity. It also relies on the looks, gestures, and implied conversations among the foot soldiers who lead the sultan, the figure descending the stairs of the classical building in the middle ground on the right, the birds that fly overhead, a riderless horse being hurried across the path of the procession, and the winding path of countless figures inching toward the Fatih Mosque on the left horizon.[27] Combined, all these elements support the image's autoptic claims and imbue it with a temporal dimension.

Yet despite Coecke van Aelst's effort to present a believable record of his visual encounter with the sultan, the picture betrays a visible tension between the static, sculptural nature of the Ottoman ruler—which epitomizes his authority in the image—and the dynamic logic of the procession. One of the reasons behind giving automata that represented Ottomans to Ottoman officials was the objects' ability to resolve this conflict. The automata overcome the tension between the rigid figuration of the sultan and the dynamic logic of the procession, because they are able to animate the representation without forfeiting any of the sultan's stateliness. Furthermore, unlike two-dimensional images of the sultan performing Ottoman pomp that exist outside of experienced time and space, the automata populate and participate in the viewer's experience of space and movement of time, as do processions. To be clear, I am not suggesting that the difference between the mimetic veracity of the print and the automata is one of degree; I am suggesting that they are different in kind. Whereas the print depends on an accumulation of visual detail to strengthen its claim to realism, the automaton relies on its three-dimensionality and its ability to move itself for a duration of time.

Surprisingly, despite their paradoxical ability to re-create Ottoman processions while maintaining an unchanging image of the sultan, the automata offer little information about the figures' particular physiognomy or psychology. Thus the automata are references that only point to a specific target but do not fully describe it. I will later turn to why these stereotypical images of the sultan may have been considered appropriate gifts. First, let us consider how these generic images of Ottoman rulers corresponded to other contemporaneous and positive representations of the sultan in Christendom.

The True Image of the Sultan

Representations of Ottoman rulers produced at the same time as the automata include prints published and widely dispersed in the Holy Roman Empire and in Europe—an array of portrayals that aimed to identify, catalog, and describe the sultan.[28] Bronwen Wilson has traced the publication of both costume and portrait books featuring Ottoman rulers and military leaders, calling attention to their diverse mechanisms for conveying difference and sameness to a European audience.[29] In Venice, costume books such as Nicolas de Nicolay's *The Navigations, Peregrinations, and Voyages to Turkey* (1567, fig. 40) influenced how Ottomans were represented in a variety of contexts. For instance, Francesco Sansovino's *Informatione* (1570), a text that urged the Venetian Senate to wage war against its Muslim neighbors, published woodcuts copied after the engravings in Nicolay's treatise. These illustrations foregrounded sartorial display—elaborate headdresses, colorful caftans, and beard lengths—to distinguish the Ottoman sultan. Similarly, popular "portraits" of Ottoman rulers, such as the anonymously published *Selin* (1580, fig. 41), stop short of portraying physiognomic or psychological individuality. Generic features are brought together with turbans and blandly ornamented garments to convey to the viewer a spurious image of Sultan Selim II (r. 1566–74). Only the label, which reads "Selin imperator de Turchi," clearly identifies the schematic portrayal as a specific personage.

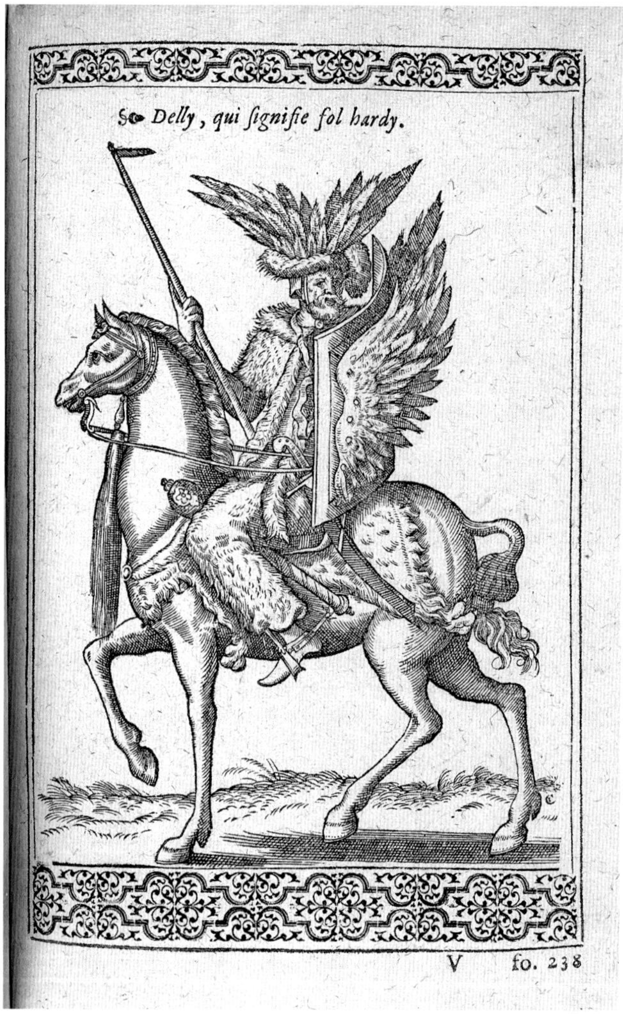

40. Nicolas de Nicolay, *Delly*, from *Les quatre premiers livres des navigations*, . . . 1568.

Images of the sultans in "portrait books" stand in stark contrast to those in costume books, for these were charged with realistic detail. Little of this detail, however, was grounded in actual knowledge about the figures represented. For instance, many of the portraits published in Pietro Bertelli's *The Lives of Turkish Emperors Accompanied by Portraits* (1599, fig. 42) portray Ottoman rulers who had long been dead. Bertelli's portrait of Sultan Mehmet II (r. 1444–46 and 1451–81) offers far more visual information about the deceased Sultan than does the image of Selim II, which is generic in the extreme. Mehmet's portrait is distinguished by portrait-like physiognomic detail. The skin of the sultan's face is aged. It puffs and wrinkles below his eyes and is furrowed along his brow. His illuminated roman nose arcs downward, calling attention to his thin, parted lips. A thick neck with loose jowls seamlessly blends into the sultan's sloping shoulders. Mehmet daintily grips his scepter, fingers unevenly grasping his royal attribute. The body is rendered in three-quarter view displaying his robe, the drapery of which lies randomly, bunched at his right elbow, falling in irregular folds down his breast. This detail gives the impression that this is an image of a specific historical figure.

In comparing these two modes of representing Ottoman sultans, we might conclude that the automata given as tribute function like the stereotypical image of Selim. The three-dimensional figures' lack of a secure identity would surely bolster such a claim. To equate the sultans on the automata (and their lack of particularity) with images like that of Selim II, though, would be to forget about their intended audience:

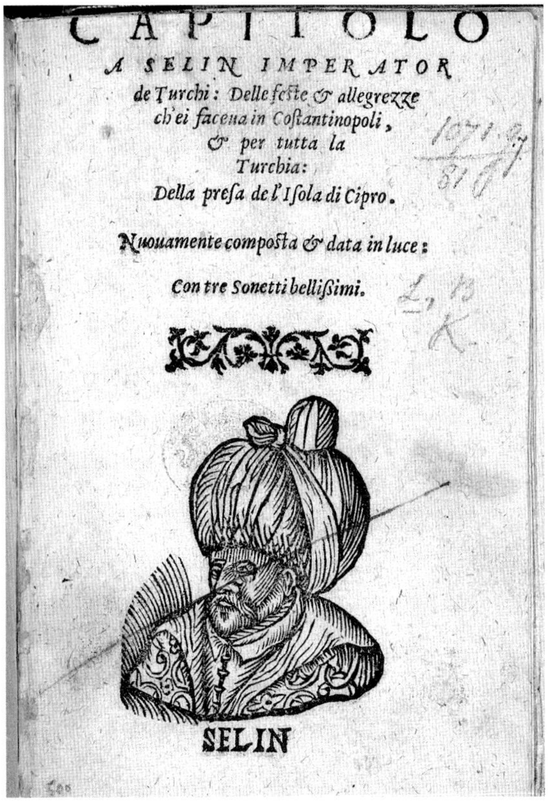

41. *Selin*, from *Capitolo a Selin imperator de Turchi . . .*, 1580.

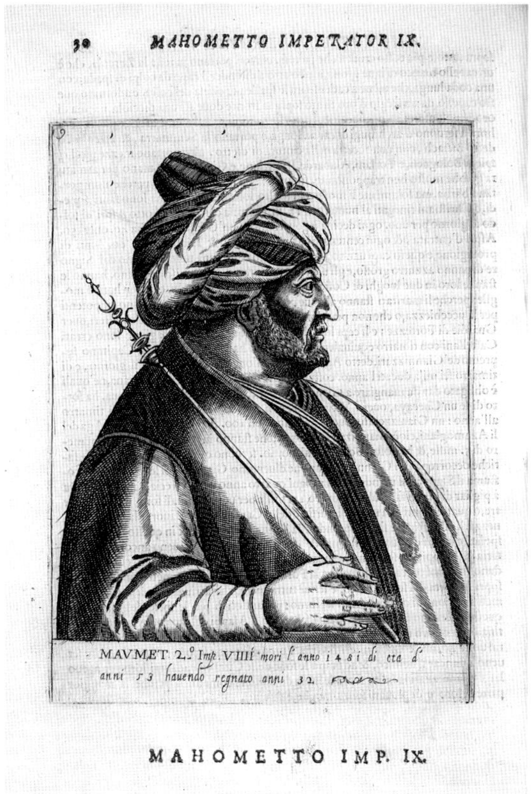

42. Pietro Bertelli, *Mahometto Imp. IX*, from *Vite de gl' imperatori de Turchi . . .*, 1599.

Ottoman rulers. Whereas stereotypical images of Ottoman sultans in Christendom were a result of a European tradition that produced generalized images of historical rulers, with detailed portraits scarce or altogether absent, the automata, in their evocation of Ottoman processions, re-present to the sultan and other Ottoman officials how the sultan represented himself to his subjects and foreign diplomats. Furthermore, the lack of individuality the automata betray resonates with how Ottoman artists represented high-ranking Ottoman officials. As Esin Atil explains, "all representations of [Ottoman] rulers were executed from memory and based on accepted models of an ideal type."[30] These iconographic conventions can be seen in a number of Ottoman portraits, for example the anonymously produced *Portrait of Sultan Ahmet I* (fig. 43). The way the automata approach physiognomy suggests that the ruler they animate is more than merely a historical figure: the figure is an ideal type that exceeds space and time. At one level, then, the automata communicated the Holy Roman Emperors' reverence for the sultan and their acceptance of his transcendental authority. Yet at another level, the three-dimensional and animated representation of the sultan afforded the rulers of the Holy Roman Empire the opportunity to demonstrate their empire's technological acumen.

The Türkenverehrung

By the time the Türkenverehrung was negotiated in 1547, the Ottoman Empire had long been

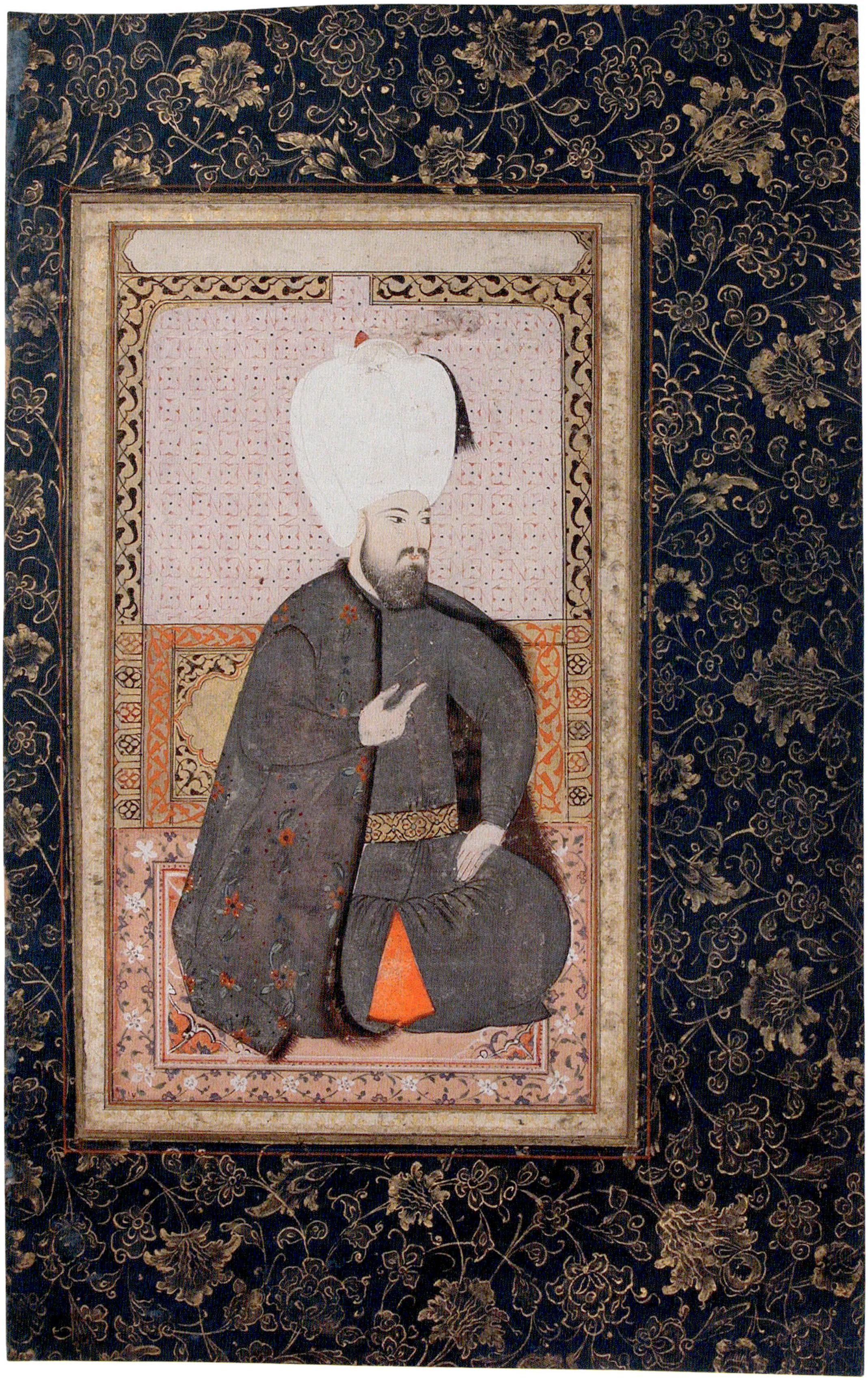

considered an existential threat to European Christendom. The fall of Constantinople in 1453 to Sultan Mehmet II (1432–1481) made clear the Ottoman territorial design on the former Roman Empire, and the titles taken up by subsequent sultans—Basileus, Padishah, and Keyser—revealed their desire to be Muslim emperors. European sovereigns and the pope recognized the tremendous stakes expeditiously, and those rulers who possessed land in the Danube valley understood immediately that they were charged with stopping the Ottoman advance. The Ottomans had reached Habsburg holdings by the late fifteenth century. In 1478 raiders interrupted a church procession in Reifnitz near Lake Wörth in Carinthia. Fear of the Turk permeated the hearts and minds of Christians in central Europe. Hundreds of pamphlets were published and numerous folksongs were written recounting and elaborating on the Turks' brutality on the battlefield and their treatment of Christian prisoners. Bells summoning Christians to pray for delivery from an Ottoman siege (*Türkenglocken*) rang out three times a day in some central European villages and towns. As Holy Roman Emperor, Maximilian I was convinced that it fell to him to eradicate the Turkish threat. Discussions of launching a new crusade were prominent at his court and in communiqués with Rome. Maximilian went so far as to reinterpret an epigram by his father Frederick III (1415–1493). Conventionally the epigraph A.E.I.O.U. was taken to mean *Austria Erit in Orbe Ultima* (Austria will triumph over the world), but Maximilian took it to mean *Austria Electa Imperatorium Ottomanicum Vincet* (Austria has been chosen to vanquish the Ottoman Empire).

The spin on the epigraph may have bolstered Maximilian's resolve, but it did nothing to block Süleyman the Magnificent's campaigns. Belgrade fell to the young sultan in 1521, and in 1526 his army destroyed the Royal Hungarian army at Mohács. But he was not planning to stop there. In 1526 Süleyman had his sights set on Vienna, but weather and low troop morale forced him to abort this plan. In the eyes of Holy Roman Emperor Charles V, who was elected after the death of Maximilian, securing the eastern border of the Habsburg Empire became critical. In accordance with Habsburg practices, marriage arrangements were made and Ferdinand, Charles's brother, secured the vacant thrones of Hungary and Bohemia. A Habsburg on the Hungarian throne, though, did not thwart Süleyman from taking the cities of Buda and Pest in 1541. These successful sieges were particularly humiliating for the House of Habsburg and alarming to the rest of European Christendom, because they positioned the Turks on the southern and northern banks of the Danube. Süleyman marked the territory as his own by converting Christian cathedrals into mosques in the cities of Esztergom, Pécs, and Székesfehérvár. At this point diplomacy must have seemed a workable option.

The truce established between Sultan Süleyman and Archduke Ferdinand was the product of seventeen years of sieges and skirmishes over control of the kingdoms of Bohemia and Hungary.[31] In the end, Ferdinand was permitted to rule Habsburg Hungary and the Kingdom of Bohemia, while Süleyman controlled Ottoman Hungary and Transylvania.[32] A crucial component of this truce—and the condition on which the Habsburgs were allowed to hold sway over their portion of Hungary and Bohemia—was the annual payment of forty thousand ducats to the sultan.[33] The treaty also specified that the peace and the division of land between the two empires could be renewed every five years, as long as the Austrian princes paid the honorarium in full and on time in Istanbul.[34] The Habsburgs dutifully satisfied these Ottoman mandates in order to maintain a pacific relationship with the Ottomans, which they did until the outbreak of

43. *Portrait of Sultan Ahmet I*, ca. 1600, Metropolitan Museum of Art, New York.

the Turkish War in 1593. The agreement signed by Ferdinand I and Süleyman did not stipulate that the ducats were to be supplemented by precious objects.[35] These addenda were taken as givens in accordance with early modern diplomatic practices.[36]

In the sixteenth century there were two categories of diplomatic occasions: ceremonies and negotiations. In a ceremony, such as a marriage, baptism, coronation, or funeral, a nuncio of the sovereign principal stood in for his employer. In negotiations a procurator, often a trained jurist, voiced the interests of his principal and arranged, among other ventures, marriage alliances, terms of war, and trade agreements. Generally there were also two forms of diplomatic missions. The first was a singular mission, in which an envoy traveled to a single court to negotiate a dispute or witness a ceremony and then directly returned to the court of his employer. The second was a circular mission, in which the envoy traveled to a number of courts without returning to the principal's court for an extended period.[37] The diplomatic occasions and missions of the Türkenverehrung did not correspond to these standard forms. Rather, the Habsburg presence in Istanbul was what the diplomatic historian M. S. Anderson has termed an "Embassy of Obedience."[38] That is, the diplomats representing the Holy Roman Empire were sent to the Sublime Porte to affirm the Ottoman Empire's subjugation of the most powerful sovereign in Christendom. A further peculiarity of the Türkenverehrung was that the representatives of the Holy Roman Empire were resident ambassadors who sometimes lived in Istanbul for decades at a time. The resident imperial ambassadors had two chief functions: to ceremoniously present to the sultan and Ottoman officials the annual tribute payment and to provide the Holy Roman Emperor with intelligence and gossip issuing from the center of Ottoman power.[39] Throughout the sixteenth century the Holy Roman Empire had only one other diplomat in residence—at the papal court in Rome.[40] Diplomats sent to other major European powers, such as France, Spain, Venice, and England, remained at the courts only briefly and then returned to Vienna or Prague.

The Türkenverehrung also differed from other diplomatic engagements in that it was a unilateral venture; the Turks rarely sent ambassadors to the imperial court in Vienna.[41] The Ottoman government possessed an unshakable sense of superiority and showed itself as being uninterested in reciprocal diplomatic relations with its powerful adversary.[42] Additionally, the imperial diplomat who resided in Istanbul was kept prisoner in his own home. He was permitted to speak neither with Ottoman citizens nor with anyone attached to the sultan's court. In one of his *Turkish Letters* the famed imperial diplomat Ogier Ghiselin de Busbecq (1521–1592), who resided in Istanbul from 1554 to 1562, describes how after being locked in his residence in Istanbul for months, he crafted a battering ram out of a ceiling beam and tried to break open the locked iron gates guarding the main entrance to his residence—in vain.[43] (Of course, diplomats adept at bribery found circuitous ways to acquire intelligence pertaining to the workings of the Ottoman court and the sultan's military campaigns.) This oppressive surveillance and control of the imperial diplomat contrasts starkly with the circumstances of other imperial ambassadors, who were incorporated into the everyday lives of the courts of Christian sovereigns.[44]

The volume of tribute sent to Istanbul far exceeded Habsburg diplomatic practices of prestation. In April 1565 six silver drinking vessels were sent to the papal court in Rome.[45] In August of the same year, an embassy representing the emperor traveling to Florence was armed with only one pendant, for Cosimo I de' Medici (1519–1574).[46] Three years later, a variety of gilded silver goblets were given to an imperial envoy traveling to London to present to Queen Elizabeth I (1533–1603).[47] Compare these numbers to an account book drafted in 1556 by the military paymaster in Vienna that lists more than 340 Ottoman recipients, not including the sultan, of gold coins, drinking vessels, small clocks, or automata in that year alone.[48] Ottoman officials often requested additional gifts. For example, in 1549, Ferdinand I personally commissioned an elaborate silver vessel and armor for the pasha in Ofen, and seventeen years later Maximilian II took it upon himself to ensure that an extra clock be given to an unnamed recipient in Istanbul.[49]

The Germans were not alone in bestowing resplendent objects on the sultan and Ottoman administrators. In 1560 the chief of the Colchians delivered a brilliant ruby that functioned as a serving vessel and twenty white falcons to Süleyman's camp outside Edirne.[50] Four years earlier, the Persian ambassador had presented the sultan with "carpets of the finest texture, Babylonian tent-hangings embroidered on either side in various colors, a harness and trappings of exquisite workmanship, scimitars from Damascus adorned with jewels and shields of wonderful beauty."[51] Venetian doges and merchants also presented the Ottoman court with bolts of vibrant textiles, colorful glass, and precious and finely wrought metalwork. In 1584 the Venetian senate granted money to be spent on robes made of cloth-of-gold for the sultan's personal tutor, Hoca Sa'deddin (1536–1599), and two of the sultan's sisters. During the same year a scholar at the sultan's court named Mehmet was presented with windowpanes and glass lamps from the Venetian Republic, while the sultan, an ardent hunter, received several gyrfalcons. The Colchians, Persians, and Venetians gave gifts to sustain and cultivate mercantile relationships with the Ottoman Empire or to alleviate tensions over military or political blunders, such as when the Venetians sank Ottoman trading ships off their coast in 1523.[52] These gifts were not a form of tribute, nor were they annually presented, like the Türkenverehrung.

But how was the tribute from the Holy Roman Empire received at the Ottoman court? We know that once a year the gifts were paraded before the sultan in the third courtyard of Topkapı Palace. But we know next to nothing about how the Ottomans responded to this ceremonial display. A page from an illustrated album in Vienna by the Flemish imperial envoy Albert de Wyt is purported to document Ottoman political ritual, but it only tells us how the reception of the tribute was represented to a Habsburg audience (fig. 44). Here the sultan is seated on a covered throne and framed by an arch. The three figures kneeling before him signal his royal status, and the Ottoman crescent perched atop a rounded pediment indicates that this is his palace (according to Wyt, the sultan's palace in Istanbul is not at all unlike sixteenth-century Flemish architecture). With eyes fixed on the courtyard before him, the sultan watches the transaction unfolding in the foreground. The tribute offering, for those viewing the event, is orderly. Equipped with golden vessels that appear to be three-quarters the size of a human being, the imperial embassy is an appropriately restrained retinue. The figure closest to us, dressed in a green caftan and red stockings, hands off a golden table clock whose movements are conspicuously visible.

Wyt pictures Ottomans in a variety of activities in the courtyard of the palace. Unlike the stately interaction between the imperial and Ottoman representatives, though, these aspects of the scene appear supplemental—an ethnographic gloss. The game-playing Turks in the left middle ground and a group of conversing Turks in the left background bolster the picture's claim, written above the scene, "Vray Pourctrait de las salle de laudience du grand S[eigneu]r dedans son Pallais." They draw attention to the choreographed quality of the transfer of objects from the Holy Roman Empire to the Ottoman Empire while simultaneously producing its unpredictable background noise. Yet imbuing the image with a believability that rests on nonessential details is not the only function these figures serve. They also form a critical audience

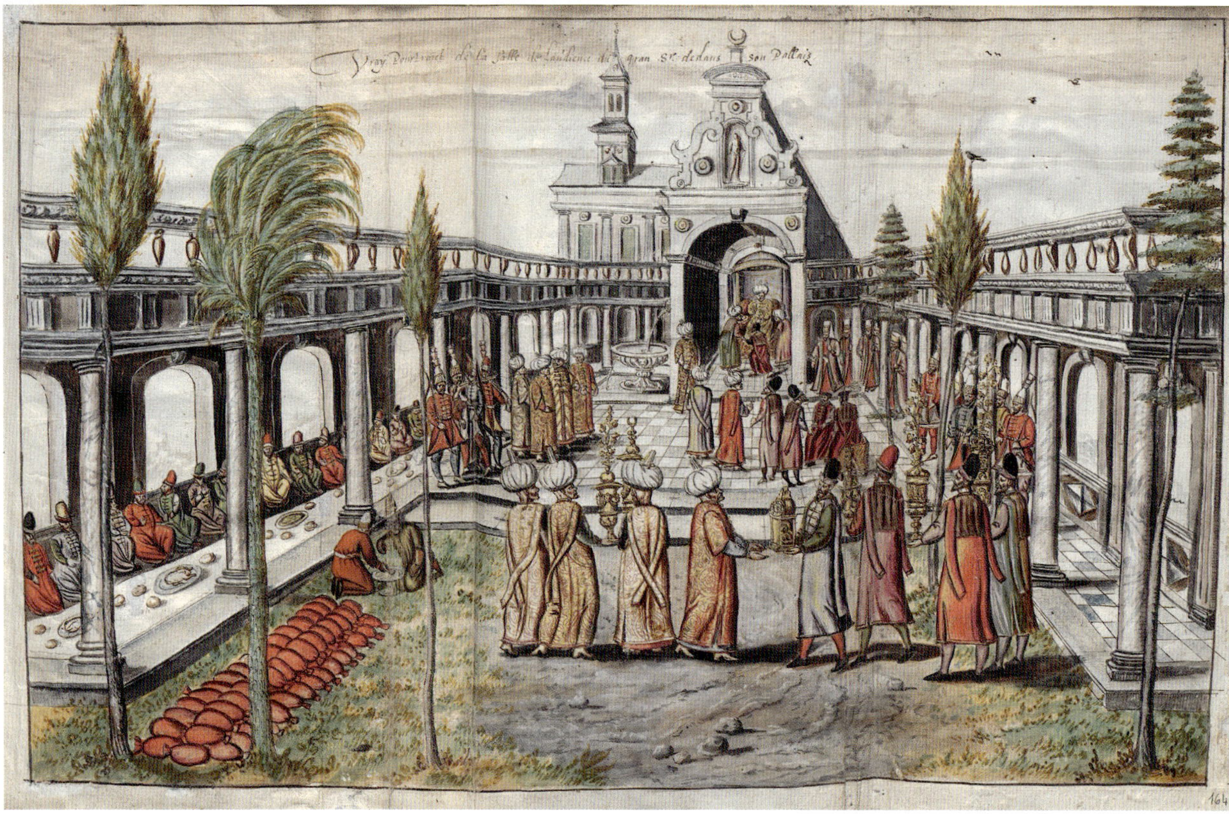

44. Albert de Wyt, *Presentation of the Tribute to the Sultan*, Cod. 3325*, fol. 164r, 1574, Österreichische Nationalbibliothek, Vienna.

audience before which the tribute is presented. Conveying that the tribute was received without hostility, they relieve any anxiety about the potential failure of the tribute. There is more. Wyt's visual account of the presentation of the tribute also communicates that the objects given as tribute were integrated into the ceremonial practices of the palace before a group of witnesses. The inclusion of the tribute into the Ottoman "ritual technology of display" made apparent, to Habsburg eyes, that the objects had accrued "publicity value."[53]

Whereas Wyt's image shows the ceremonial presentation of the tribute, a written account of the Ottoman reception recounts the objects' destiny after their public display. Imperial envoy, Lutheran preacher, iconoclast, and first translator of the Koran into German Salomon Schweigger wrote in *A New Description of the Journey from Germany to Constantinople and Jerusalem* (1578–81, printed 1608): "Although these credences and silverware are all exquisitely made, and are of much greater value than the metal, the gold and silver, this is not how these people see it, they hold it in great awe, but I am told that they melt it all down again and make coins or money from it. The beautiful clockwork that the sultan has received over the years are piled in a chamber; they are destroyed by rust and some are sold off, and he has them changed around bringing new ones to the chamber and when he has used one for some time, has it taken away, and so on."[54] Schweigger's contemptuous account of the fate of the gifts is revealing. It betrays the gamble made in assuming that objects valued in one context would be valued in another. Whereas Europeans look at the credences and silverware and see only unsurpassed workmanship, transcending the intrinsic value of the materials of which the objects are made, the Ottomans, in spite of the "awe" they experience at the sight of the objects, see them only as items to be traded or even melted down. The story changes, slightly, with the "clockwork." According to Schweigger the clocks and automata sent to the sultans were destroyed not by the forge but by neglect—neglect that was the result of the sultans' unfocused energy and fickle temperament. In other words, Schweigger's account makes clear that the Ottomans' response to the gifts did not meet his, and by extension the empire's, expectations. Indeed, attention was paid to the objects, but not the right kind of attention.

As with Wyt's image, it is impossible here to evaluate the truth of Schweigger's claims, although it is worth pointing out that during the Türkenverehrung Ottoman currency had become severely debased, and therefore it is not unlikely that most of the addenda to the tribute were melted down to bolster the value of Ottoman currency.[55] This surmise is supported by the evidence of the first surviving inventory of the Enderun Hazinesi, the Topkapı Palace Treasury, taken after 1593. Compiled in 1680, the inventory lists only a few German "clocks" and astronomical instruments.[56] If the automata were not immediately destroyed for their intrinsic value, it is possible that they were placed in the treasury among the "heathen" objects, which, according to the incomplete and sketchy inventory, encompassed paintings, playing cards, illustrated books, and objects crafted from silver.[57] Since no complete inventory or contemporary account of the Topkapı Palace Treasury dating to the second half of the sixteenth century survives, the fate of these objects will perhaps remain obscure.

What is certain is that the disparity of material exchange between the Holy Roman Emperors and Ottoman rulers during the Türkenverehrung crystallized the degree and direction of subordination between giver and receiver. Because the Ottomans never offered a counter-gift, the power dynamic in this diplomatic encounter was clearly defined, and the tribute worked to articulate it as unilateral. Yet although the automata given as tribute were not reciprocated, I suggest that the objects worked to articulate a bilateral power dynamic, and one way they did this was by presenting Habsburg rulers as masters of mimetic technology.[58] This claim is amply borne out when we take into account the challenging and at times insurmountable problems that the rulers of the Holy Roman Empire faced in procuring the automata and transporting the objects from central Europe to western Asia.

Trafficking Automata

Scheduling the manufacture and shipment of the automata was often burdensome. Generally the automata and other Turkish gifts were to be completed by late July in order to travel to Istanbul under the most favorable weather conditions.[59] But this time frame was problematic for clockmakers. On October 29, 1581, Holy Roman Emperor Rudolf II wrote to the governor of Swabia, Maximilian Ilsung, requesting that he commission a clockmaker in Augsburg to make several rings set with clockwork for the Ottoman pasha in Ofen.[60] Ilsung wrote back to the emperor apologetically stating that a master clockmaker in Augsburg (unfortunately the name is not mentioned) had told him that the rings could not be delivered until the following fall at the earliest. According to Ilsung, the clockmaker was unable to craft clockwork mechanisms during the fall, winter, and spring months because of poor lighting and frigid weather.[61] Delays in receiving the tribute from Augsburg frequently required that envoys postpone their trips to Istanbul. The imperial envoy Wenceslas Wratislaw writes in his memoirs: "We spent several months of [1591] at Vienna, waiting till the jewelry, watches, and other special presents, which our ambassador was to offer not only to the Turkish emperor, but also to his pashas and grandees, were brought from Augsburg."[62] Regular scheduling conflicts notwithstanding, automata—and other objects consisting of clockwork—were still steadily commissioned and sent to Istanbul.

Once complete, objects were brought by Augsburg merchants to Vienna by barge over the Danube. In Vienna the military paymaster recorded the objects' worth and their destination—the Ottoman Empire—and they were then handed over to the imperial envoy, who accompanied the cargo to Istanbul.[63] The envoy, servants, military protection, and wagons of food, wine, gold coin, and gifts traveled over land by caravan to Ofen.[64] This stop was crucial because the pasha in Ofen always received honoraria from the Holy Roman Emperor, and this provided the envoy an opportunity to discuss issues directly concerning the problematic Hungarian territory, which the pasha oversaw.[65] Once the gifts had been ceremoniously bestowed on the pasha, the caravan was transferred back to a series of barges—numbering about fifteen—on the Danube.[66] Having survived the torrents of the river, the barges docked at the Serbian city of Nish. From there the caravan traveled southeast, down an ancient Roman road, through the foothills of the Balkan Mountains to Sofia, the capital of Bulgaria. After the diplomatic corps took rest and sustenance, it set off southward through the Balkan valleys.[67] If the party survived the valleys—whose open spaces made it vulnerable to brigandage and extortion—it then had to traverse a rugged mountain pass known as Capi Derwent (Narrow Gates) by the Turks before descending to the ancient Macedonian city of Philippolis. From there they headed along the banks of the Maritza River to Adrianopolis, before trudging through the tulip-filled fields on the outskirts of Istanbul.

Clockwork automata had an additional disadvantage of being sensitive to changes in the weather, often requiring repair when they reached their destination. A Viennese clockmaker, Wolfgang Teissenreider, visited the pasha's court in Ofen from September 9 to December 9, 1562, to repair gifts that had been sent from Vienna. Teissenreider was unable to finish his work in the fall of 1562, and according to imperial account books he was paid for his time and travel expenses to return to Ofen from March 26 to August 15, 1563.[68] Eight years later, he traveled to Istanbul with Caspar von Minkwitz, an imperial envoy, because many of the gifts had become "seriously defective."[69] After several failed attempts to transport working clocks and automata to Istanbul, the empire made it a rule to send a clockmaker along for the voyage.[70]

Sending clockmakers to Istanbul ensured that the automata and other clocks would be operative when they were presented. It also suggests that the emperors may have wanted to transfer German artisanal and technological skills to the east.[71] Mechanical clocks were not widespread in the Ottoman Empire in the second half of the sixteenth century. The first person known to have designed and built a spring-driven clock in the Ottoman Empire was the Egyptian polymath Taqi al-Din Muhammad Ibn-ma'ruf (1526–1585). A manuscript in his hand, dated to the early 1580s and now referred to as *The Revolving Planets and the Revolving Clocks*, includes diagrams and instructions for constructing a clock with a second hand—a technological innovation that had developed in Italy during the 1560s. Apart from Taqi al-Din's achievement, there existed only very small pockets of mechanical clock production in the Ottoman Empire during the early modern period.[72] If, then, the diffusion of technology was one of the Holy Roman Emperor's desired byproducts of sending clockwork objects to the Ottoman capital, it was not entirely unsuccessful. Beginning in the last quarter of the sixteenth century a small colony of German clockmakers established itself across from the Golden Horn in Galatea. The settlement flourished and produced small watches and wall clocks until the mid-eighteenth century, when English exports of pendulum clocks and watches began to dominate the international market.[73] Although the Habsburgs may have succeeded in transplanting a miniature Augsburg to the Bosphorus, this technological colonization does not fully explain why the rulers of Christendom were determined, regardless of the time involved or the costs, to repeatedly send figurative automata that represented Turks to Istanbul. Why did the practice of gifting automata endure under such conditions?

We might conclude that the automata, in all their preciousness, were a means for Ferdinand I, Maximilian II, and Rudolf II to parade "magnificence." Yet to claim that these objects were only intended to assuage the insecurities of the Holy Roman Emperors is to assume that their gifting, circulation, and display were after-effects of a cemented social and political situation. Still, if beneath the most stringently choreographed and regulated diplomatic encounters there lies the fluid give-and-take that sustains the political relationship, why would objects exchanged in a diplomatic context somehow not participate in these subtle negotiations?

We can imagine that shortly after receiving such an object an Ottoman official would have wound it up, and thus become the spectator of his own empire's pomp. No longer an actor in the drama, the sultan, grand vizier, or pasha was engaged as an observer. This uncanny displacement may well have amplified the object's ability to communicate deference by promoting the Ottomans' sense of superiority. At the same time, by gifting objects occupying the same spatial and temporal fabric as the viewer, the Holy Roman Emperors asserted their control over such mimetic technology. These were objects, to quote the governor of Swabia Georg Ilsung, that perhaps "no one or few . . . could make."[74] To put it another way, the automata themselves staged competing and overlapping claims of power.

This last point resonates with the early modern historian Valentin Groebner's suggestion that gifts "possess seductive power, eloquence, and the capacity to transform social circumstances . . . and an effective gift is thus one that evokes ambiguity."[75] Seen in this light, the automata enable an alternative account of the power relationship between the Holy Roman and Ottoman Empires. At first glance the relationship between the empires appears resolutely asymmetrical on all levels. Yet when the focus is turned to actual diplomatic practices, we see that the power was in constant flux. And it is crucial that such subtle, almost imperceptible, negotiations were handled by objects.

Six

Metamorphosis at the Mughal Court

One of the most striking features of the history of early modern German clockwork automata is their centrifugal movement away from the Holy Roman Empire in the sixteenth and seventeenth centuries—in particular their migration eastward to courts in west, central, and east Asia. Although these objects were avidly commissioned, collected, displayed, and gifted in the Holy Roman Empire, at Catholic and Protestant courts alike, the appreciation, function, and fate of self-propelled objects when they traveled to courts beyond present-day Europe did not always mirror those of their European counterparts. Automata manufactured in the Holy Roman Empire and sent as tribute to Istanbul were hardly valued at the court of the Ottoman sultan; they were either melted down for their intrinsic value or neglected and left to decay. In China, however, German-made automata, clocks, and scientific instruments were esteemed for their mechanical ingenuity; gifts from Catholic rulers such as the Wittelsbachs and Jesuit missionaries such as Nicolas Trigault ensured that the China mission maintained a steady supply of automata.[1] The efforts to circulate automata inside and outside the Holy Roman Empire suggest that the European nobility, merchants, diplomats, and missionaries who engineered this cross-cultural translation believed automata would be valued for their precious materials, their technological acumen, and their ability to convey potent political or religious arguments, as did the nefs, the Christmas Crib Automaton, the Verkehrte Welt Automaton, and the *Sultan on Horseback*. In other words, automata distributed to courts outside of Europe embodied the notion, which was prevailing in the sixteenth and seventeenth centuries, that there was an aesthetic and social commensurability between certain powerful, independent political entities in Europe and in Asia.[2] But confidence in the ability of these objects to maintain their social and aesthetic value beyond their point of origin was not always warranted. The transcultural movement of a particular kind of automaton to India, the Diana Automaton, illustrates this point (fig. 45). It was sent to the Mughal court and absorbed into Mughal artistic production, as shown by its appearance in a Mughal painting dated to 1620 and executed at the court of the fourth emperor of the Mughal dynasty, Jahangir (1569–1627; fig. 46). The Diana Automaton offers the opportunity to explore an instance when an automaton was highly valued at a foreign court. But this is also an instance when an automaton's value and meaning were dramatically transformed within its new cultural context. Accordingly, the cross-cultural translation of the Diana Automaton affords us the occasion to consider certain curiously neglected aspects of early modern globalism—namely, the transmission and circulation of German material and visual culture to India—and to partake in discussions regarding scholarly conceptions of shifts in meaning and significance across cultures.

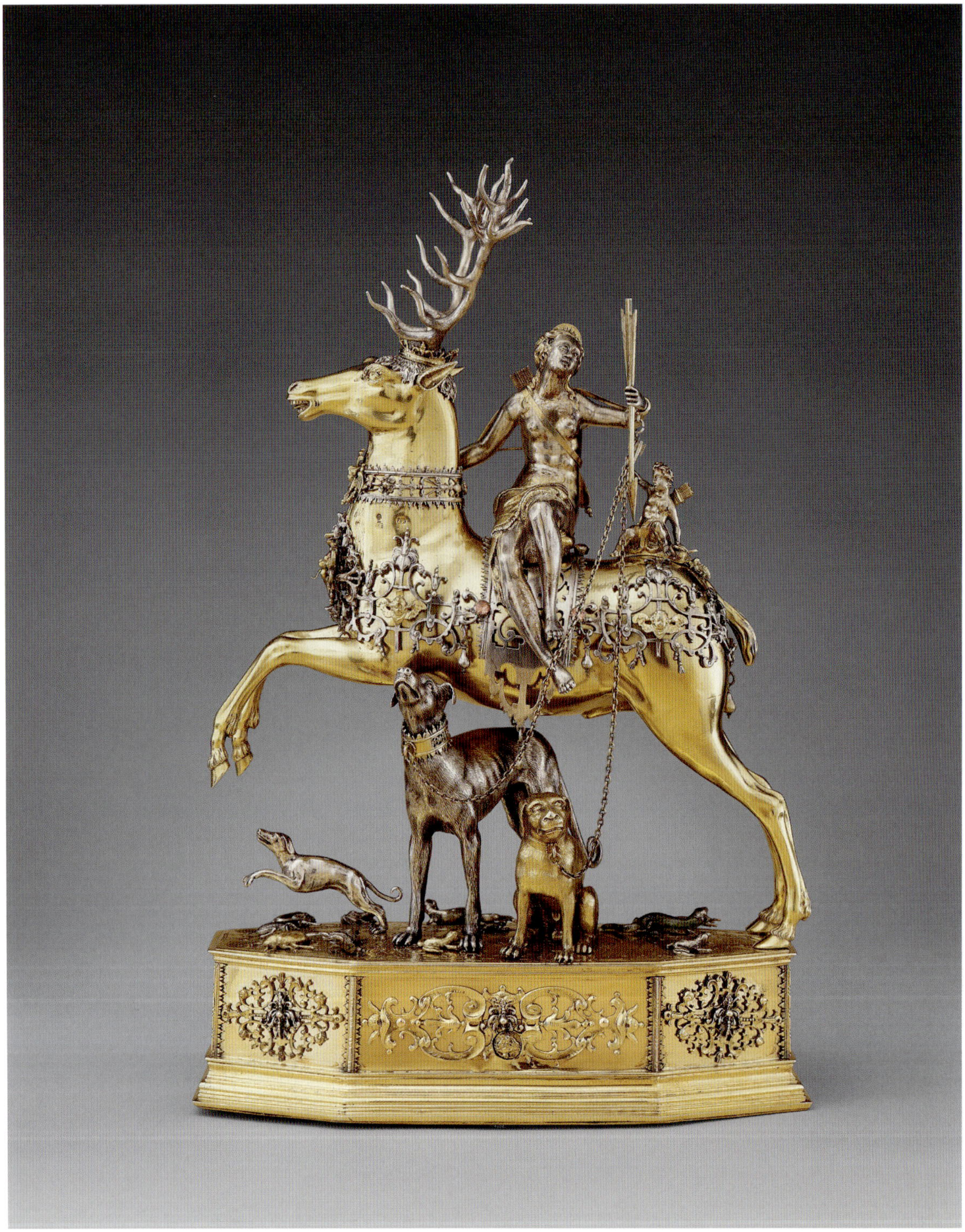

45. Joachim Friess (attributed), Diana Automaton, 1610–1620, The Metropolitan Museum of Art, New York.

46. Abu'l Hasan, *Jahangir Entertains Shah Abbas*, 1620, Freer and Sackler Galleries, Washington, D.C.

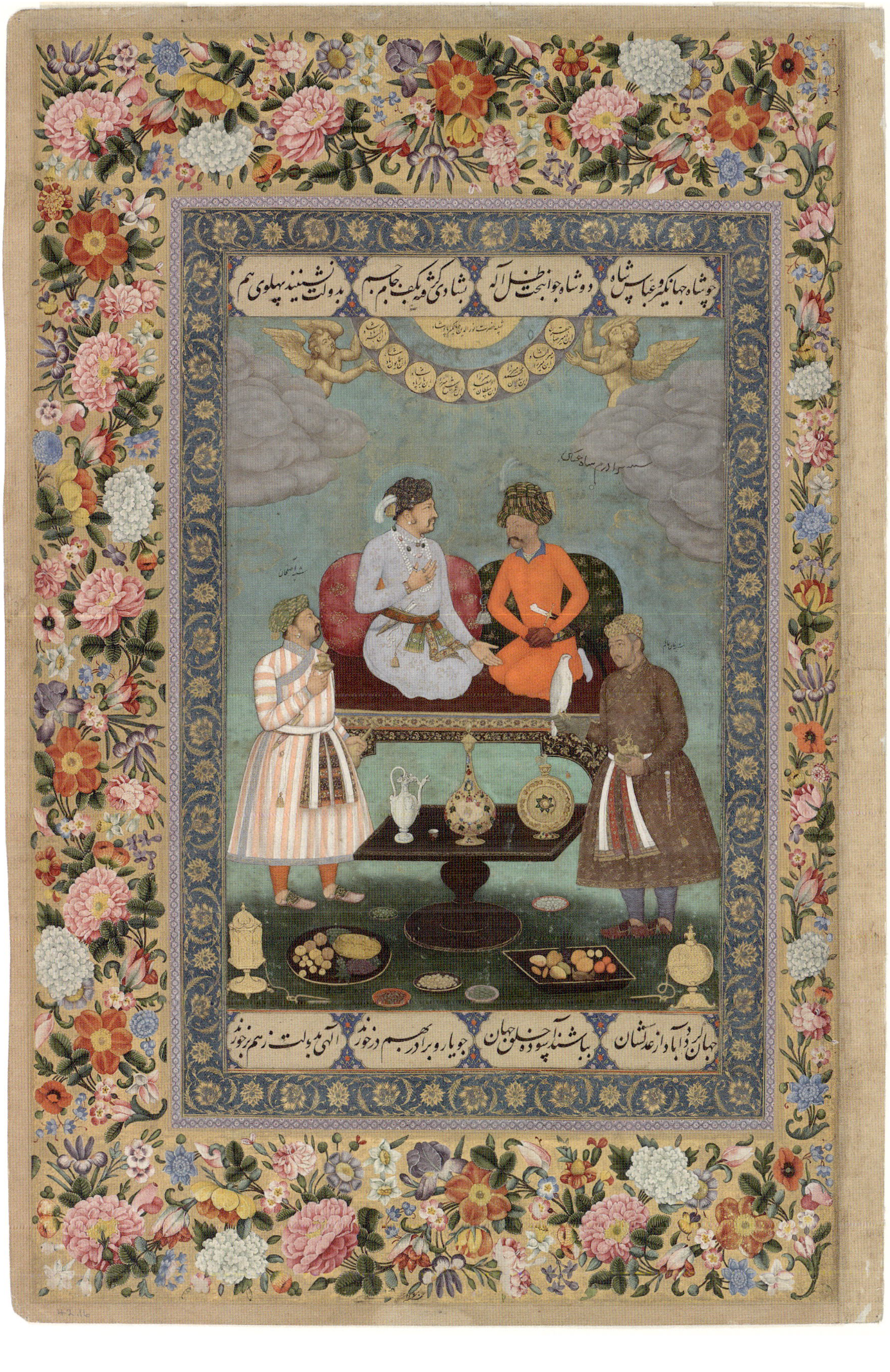

An Object of Encounter

The Diana Automaton was a type used during courtly feasts; the prototype is conventionally attributed to the Augsburg goldsmith Jakob Miller (1550–1618).[3] That twenty-five of these silver-gilt objects are extant and currently housed in museums in North America and Europe testifies to their popularity among European nobility. Holy Roman Emperor Rudolf II (1552–1612) kept a wax model of a version of the Diana Automaton at his imperial palace in Prague—presumably so that new automata could be easily and swiftly crafted for large feasts and presented as diplomatic gifts—and Amalia van Solms (1602–1675), the wife of Frederick Henry, prince of Orange and stadhouder of the Netherlands (1584–1647), owned a Diana Automaton.[4] Typically the Diana Automata were used in drinking games. The head of the stag was removed and filled with wine. Once filled, the automaton was wound and set down on the table. The person before whom the object stopped was then to drink its contents. Most of the extant Diana Automata that have been examined are programmed to move in a circle roughly 76 centimeters in diameter.[5]

The Diana Automaton displays a rather unconventional version of the myth of Diana and Actaeon. The basic plot of the myth is a tale of a chance encounter gone horribly wrong. While out stag hunting with his hounds, Actaeon, a mortal, stumbles upon the goddess Diana bathing; she is in the company of her nymphs. Surprised, the nymphs are unable to shield Diana from Actaeon's gaze. Infuriated and ashamed of her exposure, Diana transforms Actaeon into a stag, and he is immediately torn limb from limb by his canine companions, who no longer recognize him. From antiquity onward the typical moment depicted by artists—from vase painters to Titian—was the onset of Actaeon's metamorphosis from man into stag. Titian's *Diana and Actaeon* (1556–59), the earliest known example of a picture depicting Actaeon viewing the bathing scene, is an exception.[6]

The Diana Automaton departs from the myth and previous representations of it. In a representative example of the Diana Automaton type from the Metropolitan Museum of Art, a golden Actaeon, surrounded by his hounds, leaps over an array of small, cast-silver lizards and toads, while a half-clad silver Diana with gold hair, accompanied by a putto, sits sideways on its back. She braces herself on his neck with her right hand, while her left holds an arrow poised to be plunged into his rump. Here Actaeon's punishment has been transformed. Instead of a death sentence, he is to be ruled by the goddess. In the Diana Automaton, Actaeon has not just been reshaped into the prey he sought; he has also lost his freedom. Diana, armed and bare-breasted, not only rides Actaeon; she has saddled and harnessed him, and she has leashed his hounds. Moreover, unlike nearly all previous depictions, this Diana is not mortified by her nakedness. Her body is laid bare for the viewer, who becomes a voyeur like Actaeon, but the illicit nature of the gaze has been neutralized. Yet the object engages the viewer still further. Since the automaton requires that Actaeon's head be removed to drink the wine it contains, the viewer becomes the agent of Actaeon's dismemberment. All these qualities of the object—its role amplifying the viewer's enjoyment by plying him or her at feasts; its departure from the standard myth; its display of the naked body of the goddess; its placement of the viewer in the role of Actaeon's destroyer; and most of all, its ability to move of its own accord—begin to explain how the automaton accrued social and aesthetic value at courts across the Holy Roman Empire and northern Europe.

When we look farther afield, to Mughal India and the court of the fourth Mughal emperor, Jahangir, who reigned from 1605 to 1627, we find evidence that this type of automaton held appeal there as well. Again, clear evidence of the presence of a Diana Automaton at the Mughal court is offered by an imperial Mughal painting executed around 1620 by the artist Abu'l Hasan.[7] The primary subject of the painting is an encounter between Emperor Jahangir, pictured at left, and the Safavid ruler Shah Abbas I (1571–1629), at right. Now in the Freer Gallery of Art, this painting is usually titled *Jahangir Entertains Shah Abbas*.[8] Below the two potentates, seated on a divan, are Jahangir's faithful attendants. Asaf Khan, Jahangir's brother-in-law and one of his principal advisors, stands before the emperor, and Khan Alam, Jahangir's chief falconer and ambassador to the Safavid court, is situated before Abbas. Also in evidence is an array of lavish goods exogenous to the Mughal court. In addition, there is one extra-Mughal object not exhibited among the others in the foreground. This is the object conspicuously displayed in Khan Alam's left hand, a Diana Automaton (fig. 47). How did the Diana Automaton reach Mughal India? In order to begin to trace its metamorphosis from European courtly artifact to emblem of Safavid-Mughal diplomatic exchange, we must look to the longer history of Indian-German trade encounters.

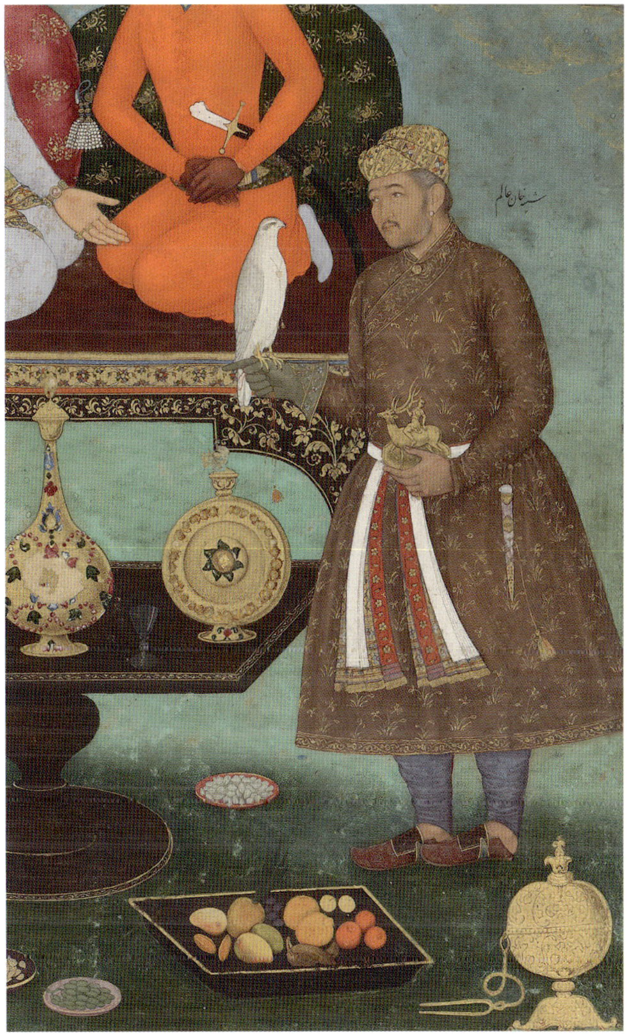

47. Abu'l Hasan, *Jahangir Entertains Shah Abbas* (detail), 1620, Freer and Sackler Galleries, Washington, D.C.

The Germans and India

It is important to explore other types of German automata and German images that made their way to the Mughal court in the sixteenth and seventeenth centuries—as well as how these objects got there and who was responsible for getting them there. Although the German Holy Roman Empire did not directly participate in colonizing the subcontinent, images and objects produced in the Holy Roman Empire circulated in great numbers in India. A good deal of the scholarship on the mobility of European visual and material culture in the early modern period concentrates on how the subjugation of one political entity by another forcefully established networks linking various parts of the globe.[9] While I am not claiming that trade and colonization are mutually exclusive, the case of the Diana

Automaton alerts us to a cross-cultural encounter in which global networks were established under conditions of a greater mutual recognition of political power and territorial integrity. European objects that appeared at the Mughal court were not sent there to replace objects or images already in use in Mughal culture; as Serge Gruzinski has shown, this sort of replacement was characteristic of European visual and material culture in post-conquest sixteenth- and seventeenth-century New Spain. Rather, European objects and images were sent to supplement the culture at the Mughal court.[10] Moreover, even though the presence of Germans in India has long been recognized and discussed by economic historians, their cultural impact in India and India's impact on the Holy Roman Empire have largely been overlooked by art historians.[11]

By the end of the fifteenth century, Indian goods, artifacts, and commodities had traveled overland to Europe for centuries.[12] But it was not until the second decade of the sixteenth century, after Vasco da Gama (1469–1524) managed to sail around the Cape of Good Hope, that a steady flow of European goods, artifacts, and commodities reached the subcontinent.[13] The opening of the cape route kindled a radical change in the circulation of not only manufactured and unmanufactured goods but humans, animals, plants, art objects, ideas, and religions as well. Yet it would be misleading to say that this change fostered contact between two worlds: it fostered the contact, collision, and intertwining of an extraordinary number of worlds, as well as stimulating new modes of approaching and engaging with the natural, the social, and the spiritual. In the Portuguese colonial enclaves along the western coast of India, German Protestant printers, Spanish clerics, Portuguese jewelers, and Italian aristocrats intermingled with Tunisian merchants, Ethiopian mercenaries, and Tamil-speaking Hindus. And in these enclaves Chinese porcelain housed Indian pepper, turkeys from the New World roamed the streets, and caskets were carved from African ivory following the model of German prints. One of the loci of this ethnic, class, religious, and intellectual diversity in India was the port of Goa.[14]

An island at the point where the Mandovi and the Zuari Rivers enter the Arabian Sea, Goa was the ideally situated administrative center of the Portuguese State of the Indies. Already by the second decade of the sixteenth century, Portuguese and other European travelers were referring to the sixteen-mile island as Golden Goa. They also employed such metaphors as "the key" and "the head" to refer to the Portuguese settlement that had within its walls the Fortaleza Palace, home to the Portuguese viceroy of India, and a Portuguese royal mint.[15] Still others baptized it "the Rome of the East," owing to the high density of churches that seemed never to cease springing up from the city's soil.[16] In his *Ethiopia Oriental*, published in 1608, the Dominican friar João dos Santos (d. 1622) offers an inventory of the things that flowed in and out of Goa: "Many ships and boats enter the port of this island coming from almost half the world. Ships from Portugal, Ethiopia, the Red Sea, Persia, Sinde, Cambay, Diu, Japan, China, the Moluccas, Malacca, Bengal, Coromandel, Ceylon and many other kingdoms and islands from these areas come here, and they are infinite. And all these ships and boats enter this port of Goa laden with many goods and riches such as gold, silver, pearls, and stones, the finest of fabrics, many silks and carpets, all the manner of spices and other medicines."[17] For European missionaries like dos Santos, Goa provided not only the perfect habitat for global trade but also a refuge from the sins of idolatry, paganism, and infidelity that ran rampant to the south along the coast and further inland.[18]

For to the dismay of the Portuguese Crown, the Portuguese ships that landed at Calicut in 1498 did not find a land providentially peopled with the lost Christians of the Ethiopian king Prester John, who they hoped would join Portugal in reconquering Jerusalem. Instead they encountered, according to King Manuel of Portugal (1469–1521), an offbeat Christian sect, whose members built chapels with walls covered in representations of "our lady" and "saints"—representations that troublingly possessed ferocious teeth and four, sometimes even five, arms. It soon became clear that these were not Christians who wavered in their beliefs but idol-worshiping pagans, effectively dashing the royal house's messianic dreams.[19] As a result, India joined Africa and the Americas as a Christian frontier, with the Christianization of its inhabitants being of the utmost importance for the Catholic Church. Goa, and the coastal Portuguese enclaves of Diu, Daman, Bassein, Thana, and Chaul, were outposts for Dominican, Augustinian, and later Jesuit missions. But it was Goa in particular where Catholic missionaries were supplied with the European texts, curiosities, and works of art needed to "harvest the heathen."[20]

It is reasonable to suspect that Goa's reputation as an effervescent site of trade and an outpost for Catholic missionaries prompted the Mughal emperor Akbar (1542–1605) to write to Portuguese authorities there in 1578 and request that "two learned priests" and the "books of the Law and Gospel" be sent to his court in Fatehpur Sikri.[21] Akbar's overture was in character: he was famously impervious to absolute religious boundaries. A Muslim himself, Akbar initiated a series of interfaith debates in 1575. He also practiced Zoroastrianism by establishing a sacred fire in his palace and prostrating himself before the sun. He dabbled in Jainism by renouncing all forms of hunting and removing meat from his diet, and he married a Hindu princess from Amber, who bore his son Prince Salim, the future Emperor Jahangir.[22] The Portuguese to whom he wrote did not, it seems, understand that they were being invited to join an active dialogue among representatives of numerous faiths. Hopeful that the Great Mughal was on the verge of a full-fledged conversion, the Jesuits responded eagerly. Akbar's request required that the missionaries leave the Christianized coast and head inland. Such a task was reserved for only the most academically accomplished, linguistically capable, and indefatigable members of the Society of Jesus.

Three Jesuit missions were sent to the Mughal court during Akbar's reign—in 1580, 1591, and 1595. The first of these missions was led by Rodolfo Acquaviva (1550–1583), an Italian aristocrat; Antonio Monserrate, a Spanish priest; and a Persian convert who took the name Francisco Henriques. According to the *Akbarnama*, the mission arrived with "a large caravan laden with choice goods."[23] These choice goods included an edition of Albrecht Dürer's *Small Passion* (1511, Nuremberg; fig. 48), seven bound volumes of Christophe Plantin's monumental *Polyglot Bible* (1568–73, Antwerp; fig. 49), and Abraham Ortelius's *Theatrum Orbis Terrarum* (1570, Antwerp; fig. 50). The Jesuits in attendance during their presentation witnessed Akbar's deep veneration of the *Polyglot Bible*. Monserrate tells us:

> On the third of March [the Jesuits] took to the audience chamber a copy of the Holy Bible, written in four languages and bound in seven volumes; this they showed the King. In the presence of his great nobles and religious leaders Akbar thereupon most devoutly not only kissed the Bible, but placed it on his head. He then asked in which volume the Gospel was to be found. When he was shown the right volume, he showed yet more marked reverence to it. He then told the priests to come with their Bible into his own private room, where he opened the volumes once again with great reverence and joy. He shut them up again very carefully, and deposited them in a beautiful book case, worthy of such sacred volumes, which stood in the same private room, where he spent a great deal of his spare time.[24]

Although this event is often cited as the initial Mughal contact with western images and text,

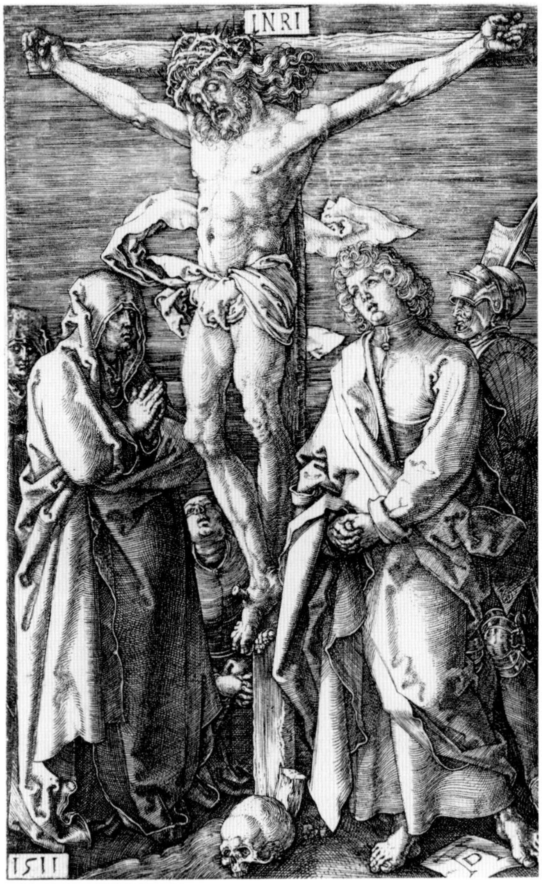

48. Albrecht Dürer, *Crucifixion*, from the *Small Passion*, 1511, Victoria and Albert Museum, London.

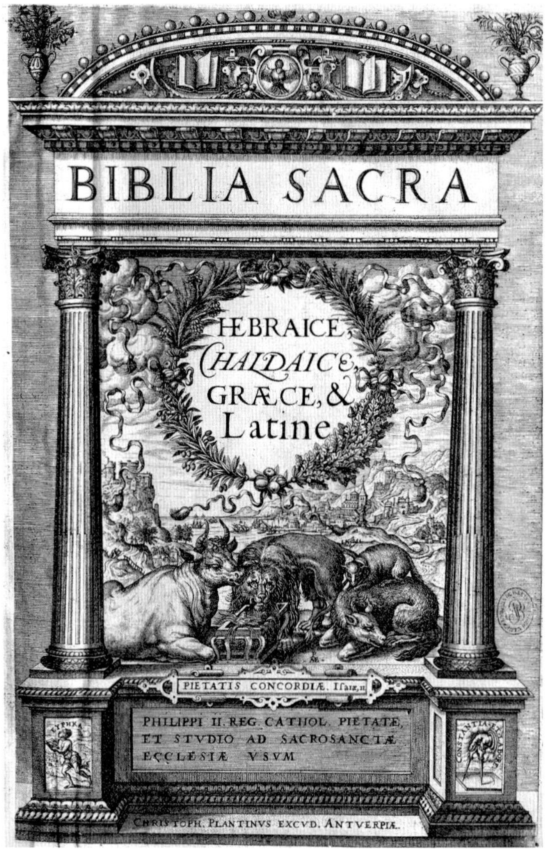

49. Christophe Plantin (publisher), frontispiece of the *Polyglot Bible*, 1568–1573, Biblioteca Nacional de España.

it was not. As Milo Beach has shown, prints by Dürer, Georg Pencz, Hans Sebald Beham, and the German Monogrammists had reached the Mughal Court before 1580.[25] Indeed, when the Jesuits presented the illustrated texts to Akbar, they may have experienced something akin to what James Clifford has called the Squanto effect, named for the Patuxet Indian Squanto (ca. 1580–1622), who welcomed the Pilgrims when they landed at Plymouth in 1620, speaking fluent English. Squanto was able to do so because he had spent several years in England and Spain after being captured by the Englishman Thomas Hunt (1585–1622). The Pilgrims' narrative of discovery ends on the note, "We thought we had discovered them, but they already knew about us."[26] If the Jesuits did experience this kind of reflexivity—if they did encounter signs that the *Polyglot Bible* was anything short of a novel event or revelation—they omitted it from their account. After all, they had not endured great suffering in their travels halfway around the world to convert an infidel already familiar with the word and images of the Lord.

That European images reached the Mughal Empire during the sixteenth century is not at all

50. Abraham Ortelius, frontispiece of the *Theatrum Orbis Terrarum*, 1570.

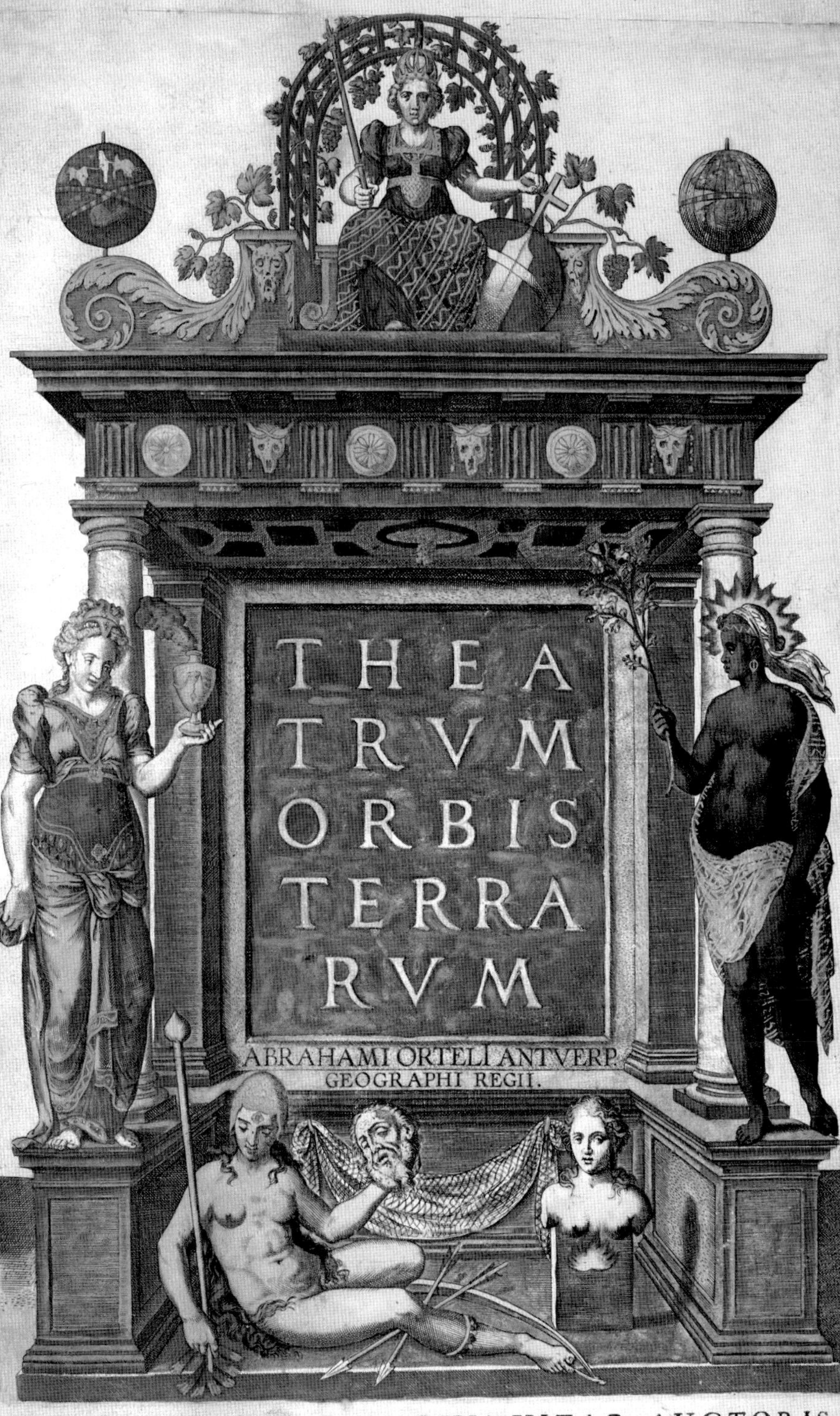

TABVLIS ALIQVOT NOVIS VITAQ. AVCTORIS
ILLVSTRATVM. EDITIO VLTIMA.

ANTVERPIÆ,

APVD IOANNEM BAPT. VRINTIVM.

ANNO CIↃ.IↃ.CIII.

surprising. Already in 1509, European printed books and printed images formed part of the spoils of war that the Portuguese obtained after a battle with the Mamluks off the coast of Diu.[27] By the 1550s western texts and images were being printed in India. A Portuguese printing press ended up in Goa at the beginning of the 1550s. Spaniards were the first to operate the press on Indian soil, beginning in 1556. By 1563 the press had been taken over by a German—Johannes von Emden—who is renowned for having published Garcia da Orta's popular *Conversations on the Simples, Drugs, and Medical Substances of India*. By the time Akbar's son, Jahangir, took the throne in 1605, the Mughal court was awash in European prints, paintings, and tapestries both religious and secular. These images were subjected to what Ebba Koch has called "interpretatio Mughalica."[28] For instance, Jahangir instructed his court painters to copy prints by Dürer, even specifying that they reproduce the effects of cross-hatching (fig. 51).[29] But it was not just Dürer whose works the court artists reproduced. A priest who accompanied the second Jesuit mission (1595) reports: "On a wall of one of the halls [in Agra] [Jahangir] had painted figures of the Pope, the Emperor, King Philip, and the Duke of Savoy, whose portraits he possessed, all on their knees adoring the holy cross, which was in their midst."[30] Many Mughal paintings that date to the first half of the seventeenth century betray the influence of European maps, like those found in Georg Braun's and Frans Hogenberg's *Cities of the World* (1572–1618, Cologne; figs. 52 and 53). And representations of Jahangir in the *jharoka*, or enthroned before his subjects and other visitors to his court, recall prints of Habsburg Holy Roman Emperors—half-figures framed by columns and covered by a baldachin (figs. 54 and 55).[31] It did not stop there. Court artists went so far as to alter European images. There are numerous paintings, dating to Jahangir's reign, in which figures from German and Flemish prints have been cut out and pasted into or around the border of

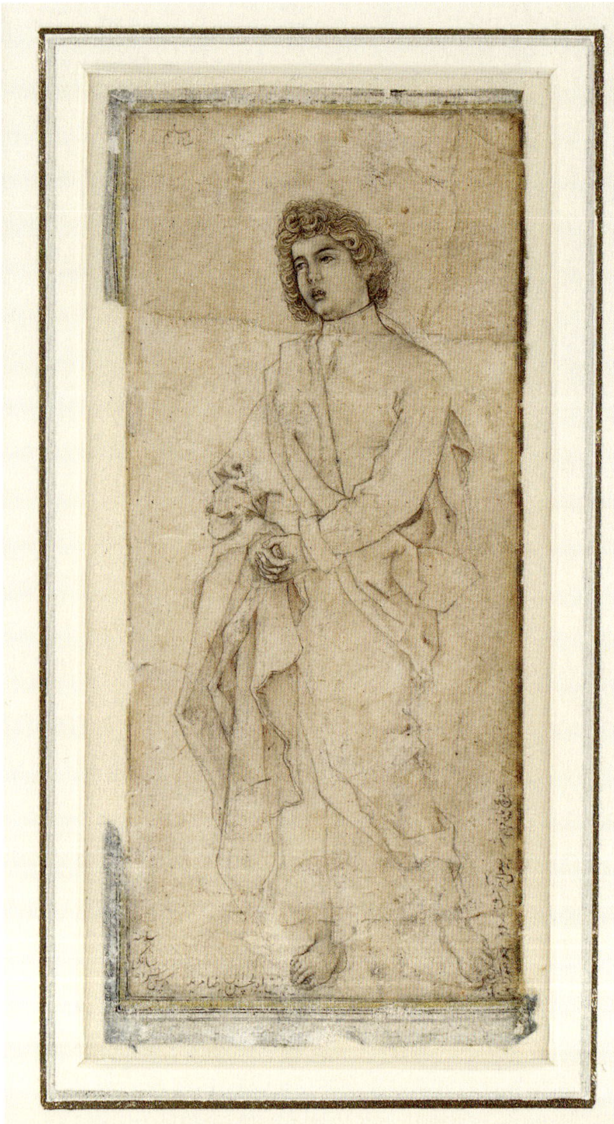

51. Abu'l Hasan, *Study of Saint John the Evangelist after Albrecht Dürer*, 1600, Ashmolean Museum, Oxford.

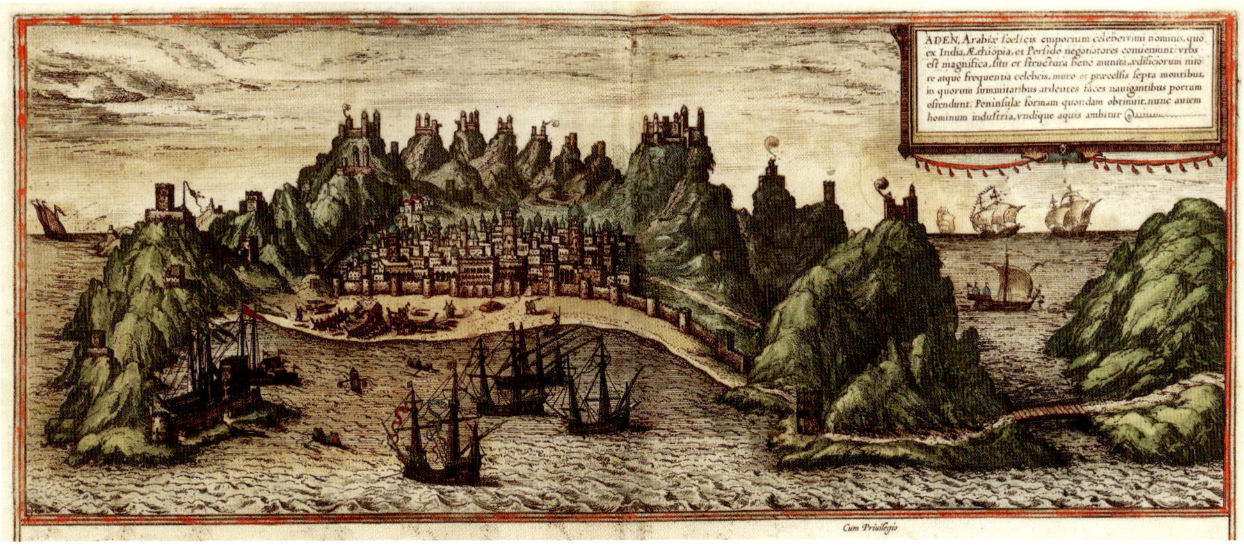

52. Georg Braun and Frans Hogenberg, *Aden*, from *Cities of the World*, 1572–1618, private collection.

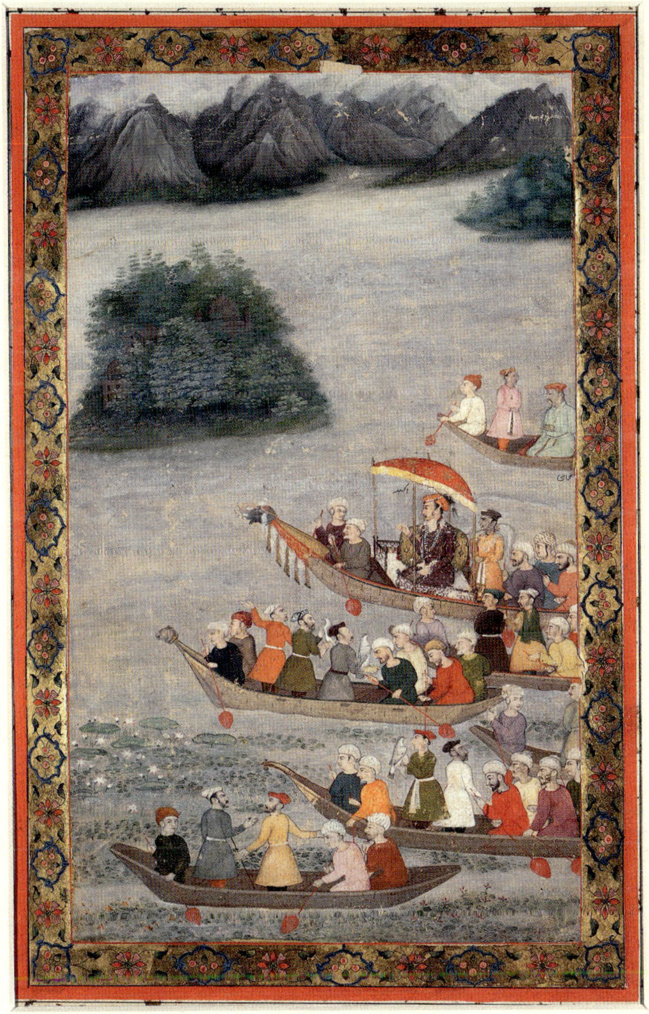

53. *Jahangir in Boat with Attendants*, 1620, British Museum, London.

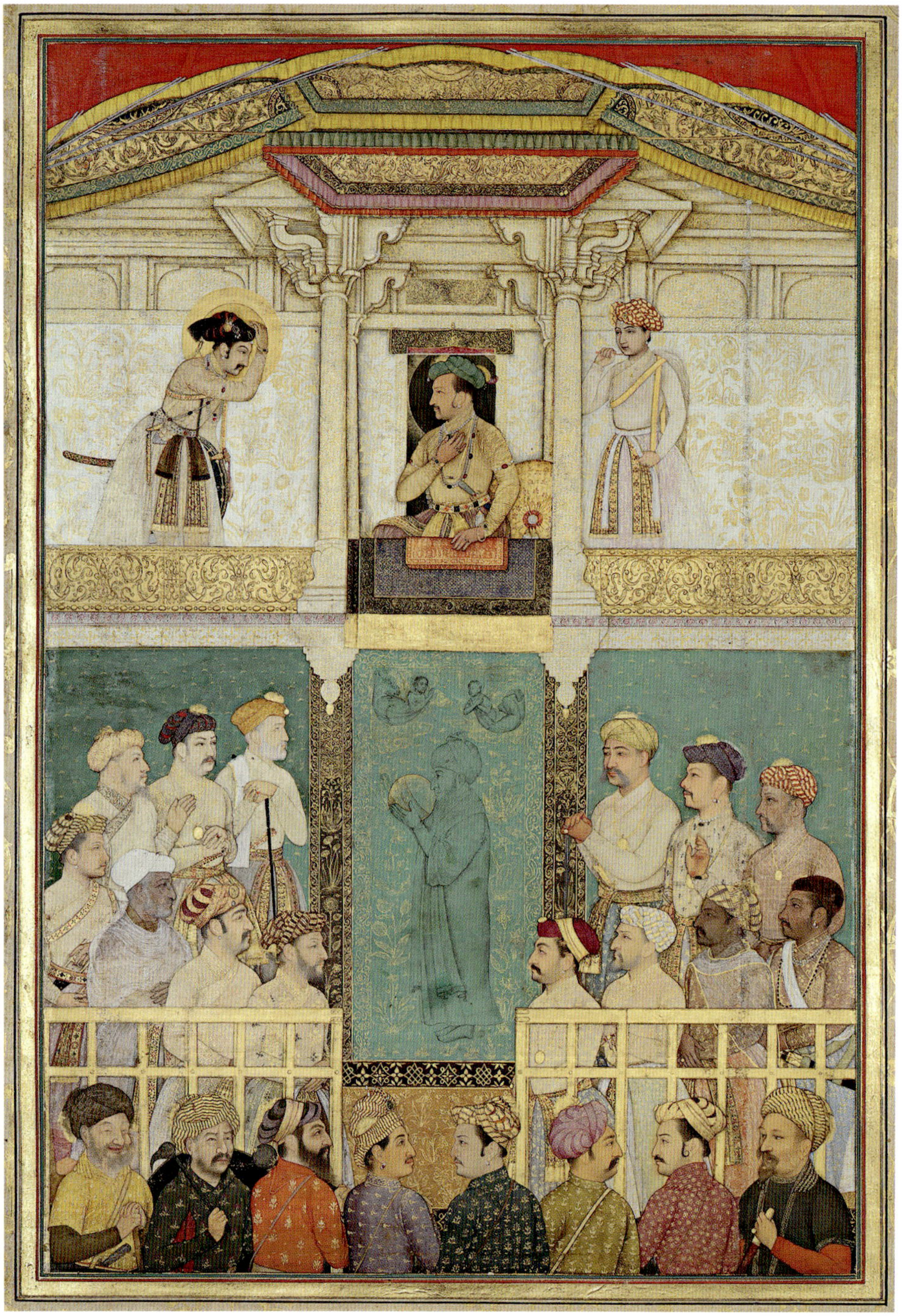

54. Abid (attributed), *Jahangir Receiving Prince Khurram, Ajmer, April 1616*, Windsor *Padshanama*, fol. 192v, 1630, Royal Collection Trust, Windsor Castle.

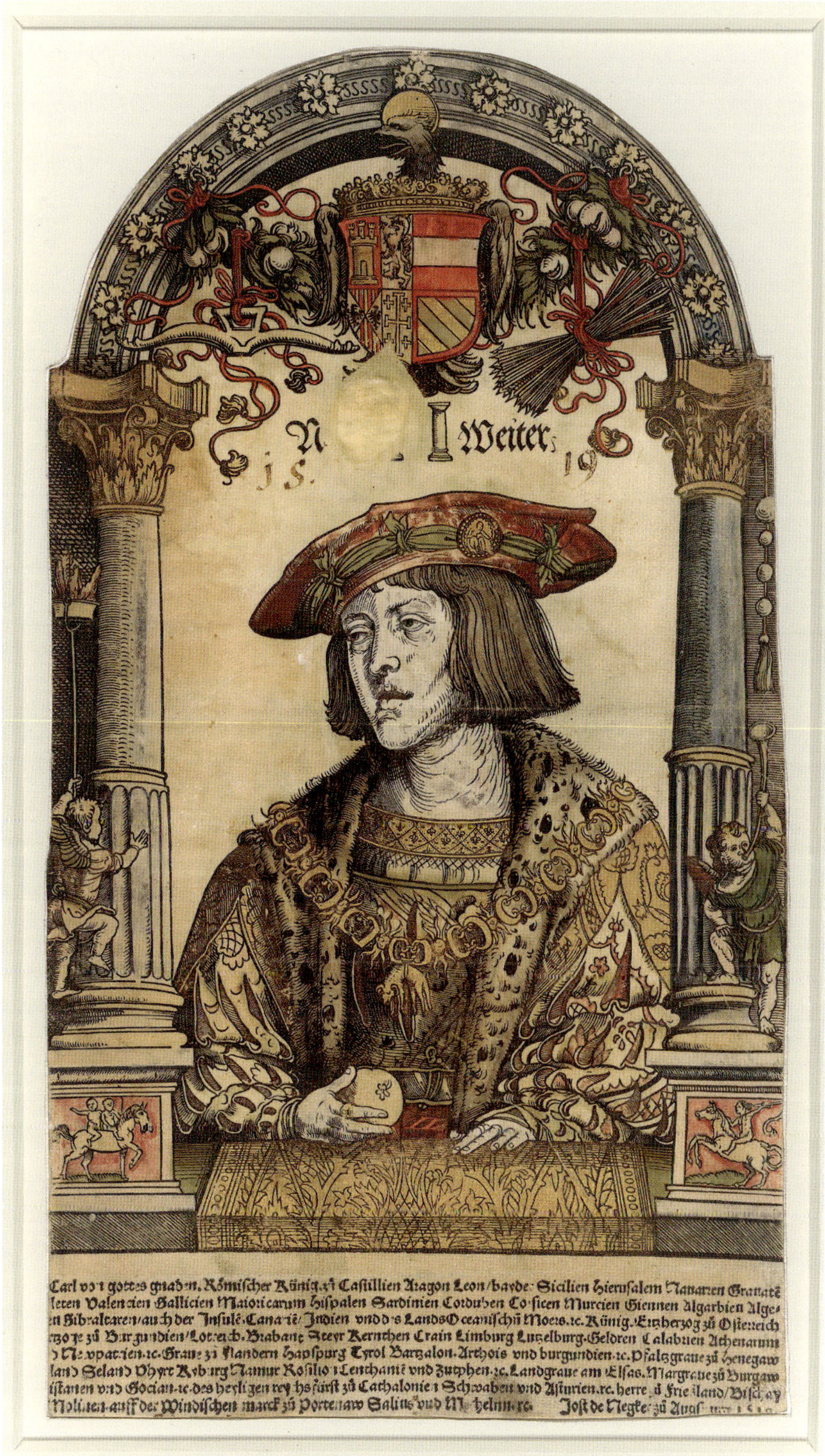

55. Hans Weiditz, *Charles V*, 1519, British Museum, London.

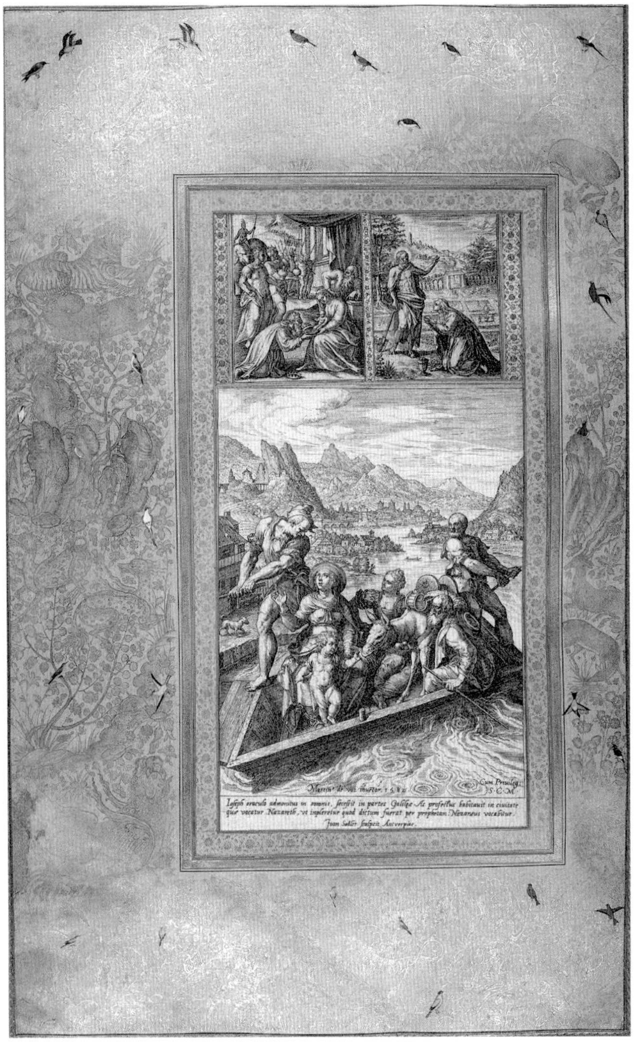

56. *The Holy Family on the Way to Nazareth*, from the *Jahangir Album*, Bl. 7b, 1605–1618, Staatsbibliothek zu Berlin, Stiftung Preussischer Kulturbesitz.

57. *The Flaying of Marsyas*, from the *Jahangir Album*, Bl. 8a, 1605–1618, Staatsbibliothek zu Berlin, Stiftung Preussischer Kulturbesitz.

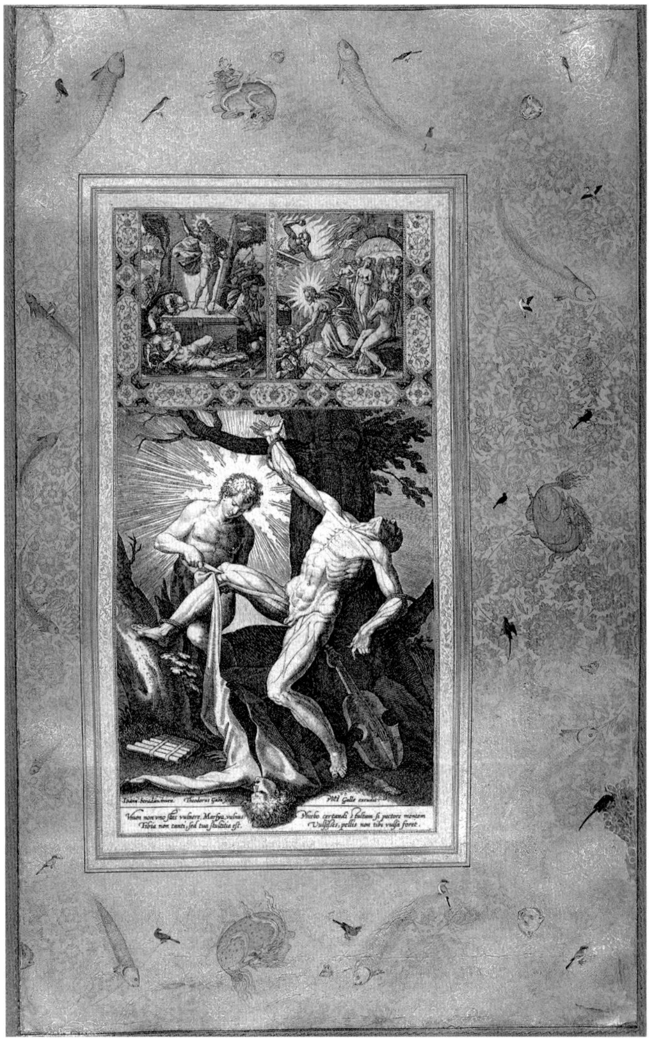

Mughal works (figs. 56 and 57).[32] These examples are striking not only as instances of early modern cross-cultural reception, adaptation, refraction, and assimilation, but also because most of the assimilated European images were not from the Iberian world, which controlled trade in India, but from northern Europe, particularly the German-speaking world.

Significant exceptions to the omission of Indian and German artistic exchange in the early modern period are the accounts of Hans Burgkmair's *Peoples of Africa and India* of 1508 and Albrecht Dürer's *Rhinoceros* of 1515.[33] Both woodcuts owe their existence to sustained German involvement in Portuguese trade in India. German mercantile families from Augsburg and Nuremberg—namely the Welsers, the Hirschvogels, and the Fuggers—funded the majority of Portuguese expeditions in India during the first three-quarters of the sixteenth century, in return for trading rights and tax exemptions on pepper, spices, and gems.[34] Portugal entered into the contracts because the Crown's funds had been severely depleted by wars in North Africa and expeditions along the western coast of Africa. As a consequence of this economic relationship, a number of German agents went to India to oversee the acquisition, storage, and shipment of commodities back to Europe. While in India these German agents developed side businesses trafficking German printed materials and curiosities to the subcontinent.[35]

Letters, news, printed materials, and luxury goods from the German-speaking world traveled overland along a route that spanned more than eight thousand kilometers (five thousand miles) from Augsburg to Venice, to Aleppo, Baghdad, and Ormuz, and then on to Goa. In the 1580s the route was extended to Macao.[36] By 1590 the entire overland communication network was organized and overseen by Ferdinand Cron of Augsburg.[37] Cron arrived in India in 1588 and settled in Goa, the cosmopolitan center of the Portuguese Indian Empire. Cron traveled there as a trade agent for the Fuggers, and by 1591, when the Fuggers terminated a trading contract with the Portuguese Crown, he had begun trading independently in European and Asian luxury goods, gems, and intelligence.[38]

While in Goa, Cron made contacts with a number of merchants who dealt directly with the Mughal court—the most important of whom was the Flemish jeweler Jacques de Coutre. In 1619 de Coutre left Goa for Agra, the seat of Jahangir's court, with gifts for the emperor and his son Prince Khurram. The objects were clearly meant to exemplify de Coutre's exotic merchandise and convey what he believed would have economic, sociopolitical, and aesthetic value at Jahangir's court. The gifts included three large cameos (unfortunately they are not described); a square piece of white jasper inlaid with lapis lazuli from Punjab; a Florentine *pietra dura* plaque representing the Archangel Michael soaring over a demon-infested hell; a pair of sheathed daggers; gold chains; and an automaton of a ship from Augsburg.[39] The mechanized vessel, which may have closely resembled the London or Écouen Nef, showcased two of early modern Europe's consummate technological advances: shipbuilding and clockmaking, forms of technology frequently involved in encounters between Europeans and non-Europeans in the sixteenth and seventeenth centuries.

De Coutre was not the first to bring an automaton to the Mughal court. Already in 1595 the Jesuit priest Jerome Xavier (1549–1617), the great-nephew of Francis Xavier (1506–1552) himself and the lead missionary of a Jesuit mission to the Mughal court, wrote to the Father Provincial in Goa and pleaded, "For the love of God, send us some good pictures on paper which depict stories such as the life of Christ our Lord, or some Saints, and some *curious* objects which we will give to the King."[40] The Father Provincial delivered. A year later, Jerome tells us that the Jesuits not only procured and displayed a Christmas

Crib during the Christmas Feast but also supplemented it with

> an ape which squirted water from its eyes and mouth, and above it a bird which sang mysteriously . . . and a globe of the world supported on the backs of two elephants . . . and above this a large portrait of King [Jahangir] which he sent us when he was a prince . . . and next to this figure was placed a large mirror at the front of the crib. . . . [At the gates] were the Angel, Gabriel, with many angels, who were accompanied by placards proclaiming *'Glory to God'* or *'Be not afraid'* in Persian. Around the Holy Infant in the crib were some sayings of the Prophets who pretold the coming of God into the World.[41]

It would take another chapter to unravel and explain this bizarre moving and resonant tableau. What is of immediate interest is that this tableau contained automata and that they, like the automaton of the galleon, came to the Mughal court via Goa. And these two instances were neither the first nor the last time automata and clockwork objects were distributed around Asia from Goa. In 1582 the Father Provincial of Goa presented to the Jesuit China mission, led by the polymath Matteo Ricci (1552–1610), a clock intended for the governors general in the southeastern provinces of Guandong and Guangxi. It is not unlikely that when the Father Provincial vested Ricci with the clocks, he also sent a clockmaker to Macao with Ricci. In the spring of 1584 Ricci notes that he was constructing clocks with "a canary islander who came from India, black-skinned."[42] Thirty-four years later Nicolas Trigault (1577–1628), the Flemish Jesuit ambassador of the China mission, was dispatched from Goa with an elaborate and complex mechanized nativity scene intended for the Chinese emperor Wan Li. Trigault described the automaton: it included three Magi in a procession before the Christ child laid to sleep in a cradle, the Virgin and Joseph rocking the cradle, an ass and an ox turning their heads toward the Christ child, God the Father making benediction, and continuously descending and ascending angels.[43]

Given Trigault's description and that we know the automaton was donated to the mission by Ferdinand Wittelsbach of Bavaria (1577–1650), the archbishop of Cologne, we can be fairly certain that the automaton sent to China was based on, if not cast from, the Christmas Crib Automaton that Sophie of Brandenburg gave to the elector Christian I on New Year's Day in 1589 and that was the subject of chapter 2.[44] First, Trigault's description of the scene and movement of the figures on the automaton sent to China corresponds *exactly* to the Christmas Crib Automaton that Sophie of Brandenburg commissioned for her husband, the elector of Saxony. Second, the Christmas Crib Automaton was crafted by the Augsburg goldsmith Hans Schlottheim, who died in 1602, but whose molds were likely left in Augsburg and used by later clockmakers and goldsmiths.[45] Third, the Wittelsbachs regularly commissioned metalwork as well as clockwork objects from Augsburg artisans for their own collection in Munich and as gifts for other European nobility.[46]

Images of Outlandish Things

How the Diana Automaton depicted in the Freer painting made its way to the Mughal court remains unclear. It is worth noting that Jahangir, like his father, cultivated a love of the arts, and we know that Jahangir appreciated gifts that displayed cleverness, labor, and artisanal dexterity. The English ambassador to Jahangir's court, Sir Thomas Roe (1581–1644), reported that the emperor was thrilled to receive from the English "a little box of crystal made by art like a ruby and cut into the stone with curious works, which was enameled and inlaid with fine gold."[47] Also, in a report in 1616 addressing the East India Company about gifts to send Jahangir, Sir Thomas Roe mentioned, "Diana this year gave great content."[48]

The principal evidence of Jahangir's appreciation for the Diana Automaton is the Freer painting of Jahangir and Abbas, whom Jahangir called "a brother." Though the two never met, they are joined here in an idealized encounter that thematizes their amity and equal standing (fig. 46).[49] Beneath a celestial hemisphere supported by two cherubs and symmetrical banks of clouds, the figures and objects are hierarchically arranged against a skylike background. A shaded area marks the ground at the foot of the divan on which Jahangir and Abbas sit and before which Asaf Khan and Khan Alam stand. Persian verses emphasizing the rulers' harmonious relations appear above and below the scene in cartouches. They read: "When the Emperor Jahangir and Shah 'Abbas, two felicitous kings and shadows of God, take to Jamshed's cup, they sit next to each other in fortune. The world flourishes through their justice, and the people of the world are at peace. Friend and brother are worthy of each other. O God, may they enjoy the fruits of each other in good fortune."[50] The Mughal and Persian emperors are joined here to "take to Jamshed's cup." This is a reference to a cup of divination, an emblem of wisdom and power.

While Jamshed's cup is not present, many other objects are, and their identities and provenances have been meticulously identified by Richard Ettinghausen and Sharon Littlefield.[51] The black pedestal table with gold edging, and the ewer and two large vessels on the table, are Italian; the transparent wine glass is Venetian; the small brown teacup adorned with a flowering plum branch is Chinese; the two golden objects in the right and left lower foreground are Portuguese; the black trays containing fruit are Japanese lacquerware. The gold cup, saucer, and vessel that Asaf Khan holds are Iranian, and Khan Alam, the chief falconer at the Mughal court, presents a falcon on his gloved right hand and cradles a German automaton of Diana, the Roman goddess of the hunt, in his left.

Although Ettinghausen describes the subject of the image as a feast between Jahangir and Abbas, we see no ritual gestures of a feast—no eating, no drinking, no music making.[52] The Freer painting is a fantastical image: the shared divan of Jahangir and Abbas levitates beneath gravity-defying golden cherubs. A feast for the eyes, this painting is an image that embodies the opulence of the Mughal court in the seventeenth century while instantiating courtly gift presentation and early modern cross-cultural exchange.[53]

Existing interpretations of this image explain the presence of the Diana Automaton as a reference to Jahangir's interest in exotic curiosities.[54] This idea seems supported by its presence among the other objects, all which are given special treatment. The Venetian glass is transparent. The plum branch on the diminutive Chinese cup is legible. Flowers and flourishes that adorn the two tall, golden Italian vessels are made apparent. In addition to rendering the particularities of the objects, the maker of the image has artfully arranged their display. The black Italian table, with clearly rendered gold edging, centrally placed, functions as a pedestal for these objects. To the left and right of the table, and echoing its top and base in shape and color, are the two lacquer plates filled with fruit. Laid out with food as if for the viewer, the three plates between the two lacquer dishes rhyme in shape and size. The two Portuguese liturgical objects in the bottom corners also seem to mirror each other: their bases, stems, and finials are all quite similar, and their gold chains are carefully splayed out toward the center of the image. An analogous formal relationship is created between the white ewer on the table and the white falcon perched on Khan Alam's right hand: the spout of the ewer and the beak of the bird parallel each other. Another reverberation is present between the vessel that Asaf Khan grips with his left hand and the long-necked, multicolored vessel on the table, as well as between the cup in his right hand and the Chinese teacup situated on its diagonal.

And yet, the automaton is also different from the other objects depicted in the painting. It has been appropriated and is as much a xeno-figure, or emblem of foreign origin or foreign power to the Mughal ruler, as an internal referent.[55] In the hand of Khan Alam, Jahangir's chief falconer, the automaton identifies him as master of the hunt. I suggest that it is possible to go even further, by purporting that the metamorphosis represented undergoes its own metamorphosis. At European courts the automaton was most highly valued because it was self-propelled. One need only consult a drawing of a Diana Automaton (fig. 58) that appears in an inventory of the Lobkowicz Kunstkammer to see that the wheel carriage of the object has been exaggerated to signal that the object could traverse surfaces. The anonymous author of this drawing compounded the mobility of the object by adding a piece of drapery that not only frames the goddess's face but also flutters as if tousled by a strong wind. At European courts the Diana Automaton was primarily brought out at feasts to temporarily transform guests, constructed as it was to bring about their inebriation. The depiction of the automaton in the Freer painting does not suggest that the object could move of its own accord. We see no wheels. It is not placed on a surface. Like the falcon on the ambassador's right hand, its motion is arrested. Together with the static, figurine-like bird, the automaton seems to serve the purpose here of identifying Khan Alam as chief falconer of the Mughal court.

By the beginning of the seventeenth century, Indian court artists and patrons were fluent in Greco-Roman mythological imagery presented in a mannerist vein. This is in large part due to the circulation of Dutch artists in the region during the period in question, the most renowned of whom was Cornelis Claesz Heda. Heda traveled first to Goa and then to Bijapur, where he established himself as a court painter to Sultan Ibrahim Adil Shah II. From Heda's many surviving letters written between 1610 and 1619, we know that the sultan fancied paintings depicting the sensual exploits of Venus, Bacchus, and Cupid.[56] Lesser-known Dutch artists have been identified at the Mughal court during the same period, although according to Sanjay Subrahmanyam, their activity has proved more difficult to trace.[57]

If, on the one hand, the depiction of the automaton in the Freer painting registers a conceptual cognate across Germano-Indian cultural spheres—Diana as a signifier of the hunt—on the other hand, omissions and additions—a lack of equivalence—characterize this appropriation.[58] In the burgeoning literature on transcultural exchange, cultural and art historians often employ hybridity as an explanatory metaphor for coming to terms with objects like the Diana Automaton that lack a stable identity as a result of their transcultural lives.[59] Hybridity is a complex concept that comes down to us from Homi Bhabha's work and has been reiterated and deployed by scholars who have contributed to the sustained vibrancy of the term for analytic purposes.[60] Echoing (though likely without knowing it) Christopher Pinney's essay on the always-already hybrid nature of European culture, the South Asian historian and philologist Sheldon Pollock occupies a radical and now widely accepted position on the ever-presence of hybridity: "There exist," he writes, "no cultural agents who are not always-already transcultured. . . . The cultural materials being transferred are already hybrid themselves; and like the transmitter the receiver culture too is

58. *Diana Automaton*, drawing from the inventory of the Lobkowicz Kunstkammer, 1650, Lobkowicz Collection, Prague.

91v.

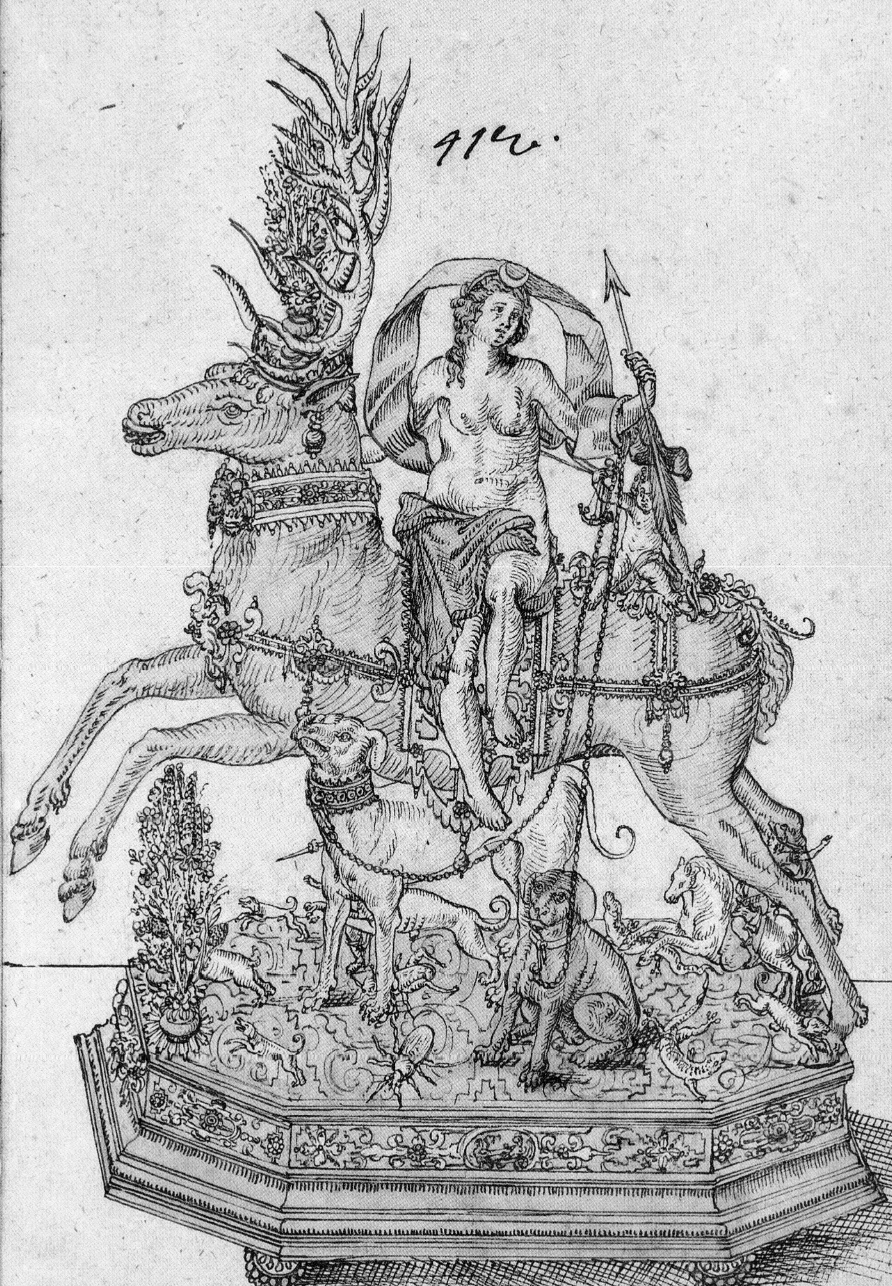

9 mark 9 ¼ Loth.

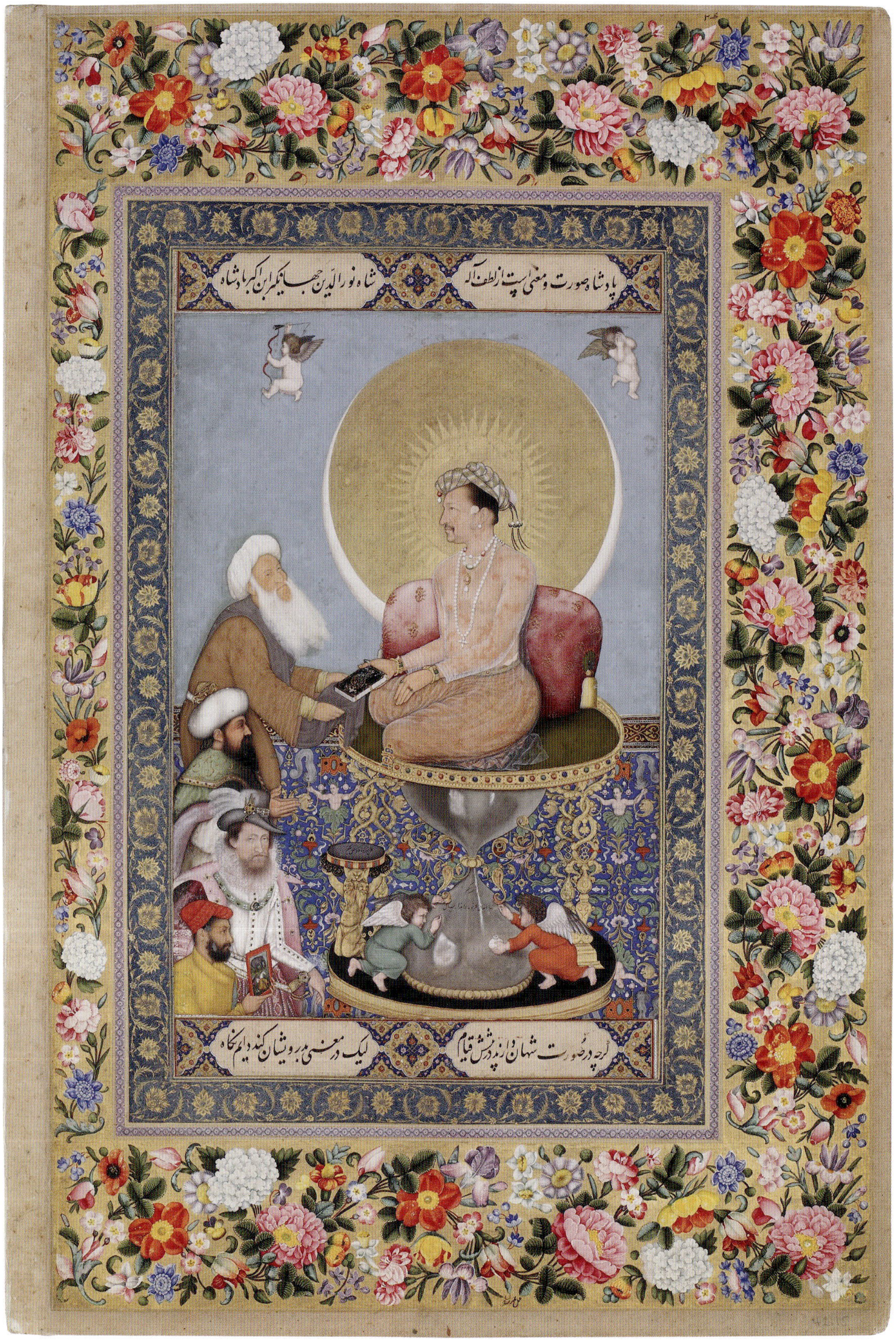
پادشاه صورت و معنی است از لطف اله
شاه نور الدین جهانگیر ابن اکبر پادشاه
گرچه در صورت شهان دارند در پیش قیام
لیک در معنی بدرویشان کند دایم نگاه

something always in process and not a thing with an essence."[61] While I agree that objects and persons who cross cultural borders confound definition and are always in process, I am less inclined to use the term *hybrid* here, where the concept of metamorphosis seems more apt. Let me make clear that although the Diana Automaton does not overtly represent metamorphosis in process, the myth of Diana and Actaeon—a myth about an encounter, about the transformation of one thing into another in space and time—embodies the essential themes of metamorphosis and as a result is useful for thinking about metamorphosis as a way of describing objects with transcultural lives.

Let us examine another instance of it. What place do exotic material things occupy in other Mughal paintings of the same period? Bichitr's painting *Jahangir Preferring a Sufi Shaikh to Kings* evidences transformations of European objects not unlike that in *Jahangir Entertains Shah Abbas*.[62] The painting, dated to 1622, reproduces a portrait of the English king James I painted by John de Critz (fig. 59), alongside representations of Jahangir, an Ottoman sultan, the Sufi shaikh Husayn Ajmiri, a self-portrait of the artist, four putti, a European textile, a European two-headed figurine, and an hourglass of European origin. The figures of the shaikh, the sultan, the king, and the artist, more diminutive than the seated Jahangir, stand and crowd the left side of the painting, squeezed between the painting's edge and the hourglass. The entire painting revolves around the hourglass, placed center stage. No longer is this object a small timepiece used to organize the temporal fabric—to divide the day into parts, as the water clocks, or clepsydras, did at the Mughal court.[63] Monumental in scale, the hourglass doubles as Jahangir's dais, his seat of power, and, by way of its trickling sand, a sign of the emperor's presence in the past, present, and future. One of the winged figures, hovering above the base of the timepiece, amplifies this motif of time, by writing out on the base a wish that Jahangir live for a thousand years.[64] The hourglass's fragility as a thing made of glass is denied in the painting; it implausibly supports the "world-holder" Jahangir, who climbed to his seat by placing underfoot the figure of a two-headed man—a tradition in accordance with Mughal "concepts to have demons, angels, or subject people carry the imperial seat."[65] Like the Diana Automaton, a marker of Khan Alam in *Jahangir Entertains Shah Abbas*, the hourglass is a transcultural object that was brought to the Mughal court from afar and has been resized and repurposed to convey something about the identity of a crucial figure in the image—in this case Jahangir's ever-presence.

59. Bichitr, *Jahangir Preferring a Sufi Shaikh to Kings*, 1620, Freer and Sackler Galleries, Washington, D.C.

As Yael Rice has shown, in Bichitr's painting Jahangir is in the midst of a visionary experience, employing his faculty of inner seeing (*Musharraf*).[66] Our visual experience of Jahangir's inner vision is anything but immediate, but two elements of the picture announce his inward state. The first is the radiant, golden halo that almost fills the center of the painting, enveloping Jahangir's torso. This halo both conveys Jahangir's inhabitation of an intermediary space between the celestial and terrestrial spheres and registers Jahangir's proximity to the light of knowledge and the light of the divine. "This zone," writes Rice, "is not only the place of dreams, but it is also . . . the *'alam al-mathal*, that is the sphere where images and imagination mingle."[67] The second element of the painting that suggests we are witnessing Jahangir turn inward is the figure's gaze. Although surrounded by figures and material objects, Jahangir looks out over the head of Husayn Ajmiri, peering into the blue blankness, unable to see any other elements of the picture.

What seems odd about *Jahangir Preferring a Sufi Shaikh to Kings* is the inwardness it celebrates and the outward expression of the collection of material signs. Here objects receive special treatment. Pearls are clearly articulated. Gems are vibrant. The painting held by Bichitr is absolutely legible. Clasps, pages, and boards on the book that Jahangir hands to the shaikh are made apparent. The glass of the hourglass is transparent. The cushion behind Jahangir molds to the sitter's back. More than any of the other figures represented, James I stands encrusted in material things. This depiction of James, who looks as if he has just stumbled onto the wrong stage, is based, as Milo Beach has shown, on a portrait attributed to the king's sergeant painter, John de Critz. Official portraits, like de Critz's *Portrait of James I* from 1605, were customarily presented to foreign rulers by English ambassadors during diplomatic missions (fig. 60).[68] That Jahangir's court painters copied English paintings is attested by the now renowned and oft-cited passage from Roe's report in which he recounts his inability to discern the difference between an English miniature and five Mughal copies of it.[69] Roe also relayed that English portraits were displayed in March 1616 at the Mughal court: "At the upper end were sett out pictures of the King of England, the Queene my lady Elisabeth, the Countesse of Sommersett and Salisbury, and of a citizen's wife of London; below them another Sir Thomas Smyth, governor of the East India Company."[70]

In Bichitr's painting this diplomatic gift, this single full-length portrait, has been markedly altered. Not only has the figure of James I been reduced in scale—he is dwarfed by not just Jahangir but the hourglass as well—it has been tucked into a composition with many figures. His appointed position in the margin of the image, beneath Husayn Ajmiri and the Ottoman sultan and slightly masked by the self-portrait of the artist, alludes to James I's inferior position, his transformation from "James I King of Great Britain, France, and Ireland, Defender of the Faith" into a lesser functionary, a mere court attendant.[71] Whereas Jahangir's human and divine status oscillates constantly in this image, the king of England has but one body that is clearly of the earthly, material world. Yet when examining this representation of James I we cannot fail to notice that it is based, like the Diana Automaton, on a single object—an official portrait—that traveled a great distance to Mughal India, nor that James I in this painting has been given a different status and a new function. The portrait through its figurative and literal translation has become something else. It is no longer treated, to quote Natasha Eaton, as a portrait that "transmit[ted] the presence of the donor to the recipient," but as an object, not unlike the ornamented European textile that functions as the backdrop to the scene.[72] To consider this re-presentation of James I's representation as a hybrid image is to disavow, implicitly and explicitly, its new identity in a Mughal context.

"When after a storm . . ."

Hybridization and metamorphosis are, to be sure, conceptually linked—but they differ in crucial ways.[73] In addition to the problematic racial freight the term *hybridity* carries, it entails the seamless fusion, or joining, of two opposing identities. In metamorphosis, however, identities remain in flux. The dynamic of metamorphosis depends on the continual oscillation, or potential for oscillation, between states. Actaeon is both man and deer, not a centaur with a divided identity. Metamorphosis and hybridization also differ in terms of their chronicity. The notion of hybridity assumes simultaneity, while metamorphosis is sequential—much like the rhetorical self-presentation of an automaton or hourglass, which unfolds sequentially.[74] Since it unfolds over time, metamorphosis encompasses multiple

60. John de Critz the Elder (attributed), *James I and VI*, ca. 1606, Dulwich Picture Gallery, London.

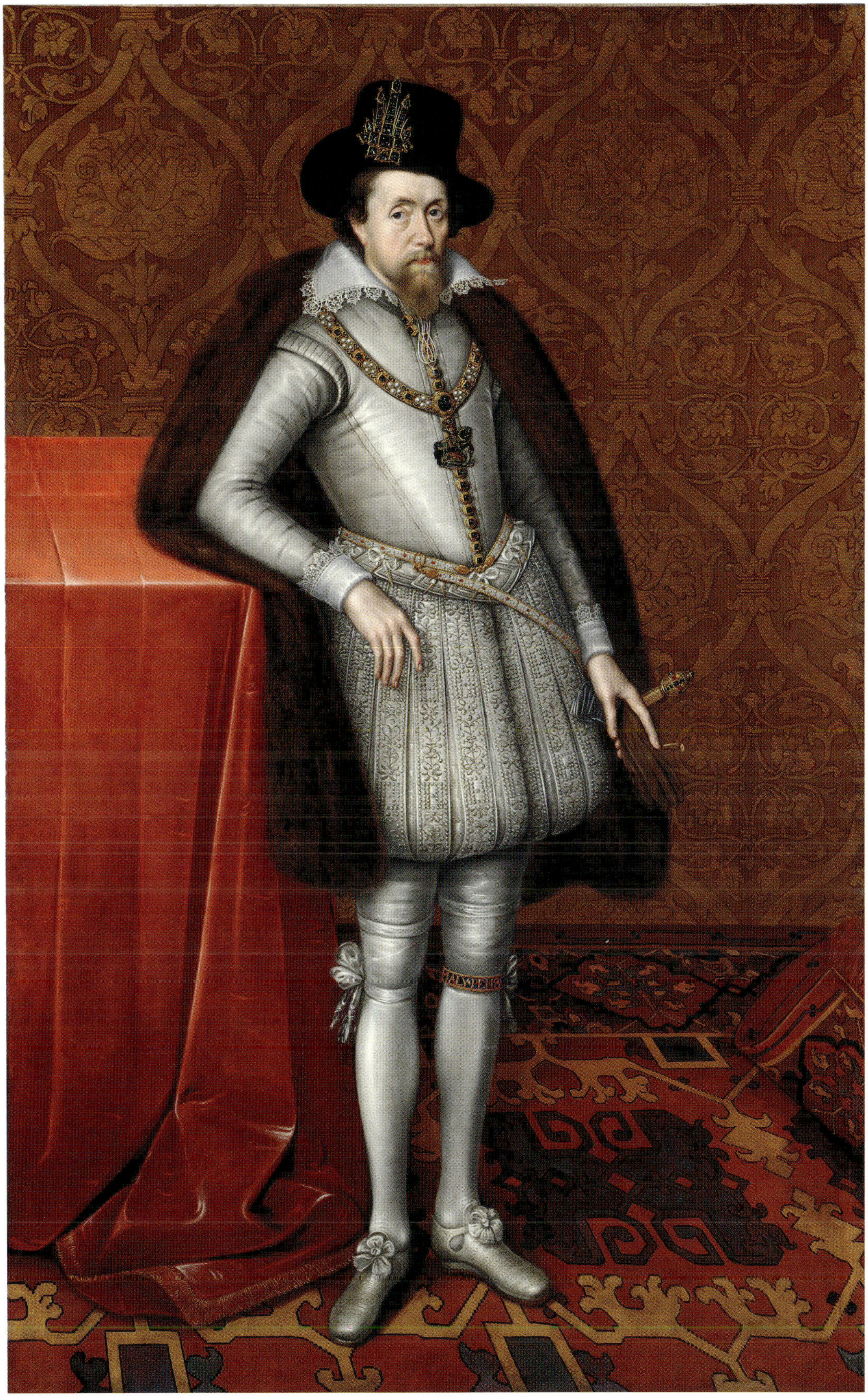

states of alteration and transformation. A passage from Ovid offers a radiant parallel. Nestled in his account of Arachne and Minerva, the poet describes the two women working swiftly and deftly at spinning, dyeing, and weaving to produce a range of colors: "As when after a storm of rain the sun's rays strike through, and a rainbow, with its huge curve, stains the wide sky, though a thousand different colors shine in it, the eye cannot detect the change from each one to the next; so like appear the adjacent colors, but the extremes are plainly different."[75]

This final chapter has spent a great deal of time worrying about the relationship between two objects: a German clockwork object that represents Diana, goddess of the hunt and agent of Actaeon's metamorphosis, an automaton produced in numbers for northern European nobility in the years around 1600, and a Mughal painting produced for Emperor Jahangir circa 1620 in which an ambassador to Jahangir's "brother" Shah Abbas holds just such an object. But what can this painting, executed in Mughal India, tell us about the function and value of the Diana Automaton beyond its point of origin in the Holy Roman Empire? We began with an examination of the Diana Automaton, and a consideration of why it accrued social and aesthetic value at northern European courts, and then moved east to discuss other German images and German automata present at the Mughal court during the reigns of Akbar and Jahangir. As we did, a few things became clear. The Holy Roman Empire was still a major player on the ever-expanding global field in the sixteenth and seventeenth centuries. In this regard, the case of the Diana Automaton amplifies the many histories that have equated early modern globalization with striking changes brought about by Iberian and later Dutch and English maritime exploration, and it reveals an aspect of India's cultural history that is not conditioned by developments concerning England, France, or Portugal. Ultimately, after an account of the circulation of objects from the Holy Roman Empire to the Mughal Empire, we turned to the Freer painting to explore how the presence of the automaton in it signals a transformation of the object's social and aesthetic value.

In addition to complicating the social life of the Diana Automaton beyond its point of origin in the Holy Roman Empire and presenting instances of German and Indian cultural exchange in the early modern period, I have had a further aim, to which I have already begun to allude. As an alternative to the dominant model for analyzing objects that crossed cultural borders—the model of hybridity—I have proposed that the inclusion of the Diana Automaton in the Freer painting itself suggests a more apt and more robust metaphor for thinking about and describing transcultural objects. Hybridity has been a useful explanatory metaphor to throw into relief the fiction of pure identity, to denote mixing and uncertainty, and to describe an aesthetic that cannot be identified as local. Accordingly, it maintains a strong conceptual hold on scholarship that deals with the transcultural flows of objects and images.[76] Yet despite its explanatory power, the hybridity metaphor has shortcomings: it collapses the temporal, spatial, and transformative dimensions of cultural contact. I propose that metamorphosis better captures what happens when, as a result of global travel and cultural re-contextualization, objects and images accrue new functions and new values.

There is one additional sense in which metamorphosis is especially useful for expanding our explanatory framework. Crucially, unlike hybridity, it does not replicate the concept that it opposes.[77] To put it another way, metamorphosis does not imply the coexistence of homogeneous, monological, and stable cultures that are incommensurable. Instead it underscores the elusiveness of unified cultural and social entities.[78] Like Ovid's stained heavens, it reflects a world that is fluid: the extremes are plainly different, but often the eye cannot detect difference among the adjacent components of the whole. The examples we have seen of the adaptation, refraction, and assimilation of German visual and material culture in Mughal India have demonstrated that the transfer and recontextualization of people and things from one culture to another are generative. While initially productive in showing how pure identity is artificial, hybridity forecloses the possibility of discerning or discovering something radically new, because it narrows our intellectual referents. Metamorphosis, by contrast, allows for a vital, mutable world of meaning.

Conclusion

In reflecting on my analysis of automata, readers will deduce that the animation of inanimate matter was neither an organized nor a cohesive endeavor in the early modern Holy Roman Empire and beyond. That assessment would be correct. Even within the Holy Roman Empire the objects presented throughout this book demonstrate the various ends to which automata were put. Between 1560 and 1610 automata were manufactured, displayed, and circulated to make claims about absolute sovereignty, Lutheran princely liberality, Protestant preaching, Habsburg superiority over the Ottomans, and cultural commensurability between East and West. Moreover, the diversity of artistic materials, forms, and iconographies that one finds in early modern automata suggests that the automaton was considered a malleable type of object by patrons and collectors. Automata were valued for their precious materials, technological acumen, capacity to delight, and ability to make visual arguments by way of their iconography and animation.

The political significance of automata forces us to think differently about their role in early modern Kunstkammern. Scholars have typically explained that the lure of automata for princely collectors resided in their ability to astonish viewers with their mimetic veracity. In this view, the precise iconography of each object is subordinate to the object's ability to mimic certain actions, such as the strumming of a lute or the brandishing of a weapon. This book has shown, however, that automata were not objects whose sole aim was to duplicate nature. Indeed, to experience a sixteenth-century automaton in motion, moving through space and time, is to be astonished at its artifice: the movements, far from fluid, are marked by quakes and shakes. And when the figures on the automata move—when they parade around an emperor, bow before the Holy Family, beat a pulpit, roll their eyes, or ride a stag—their mechanical stiffness reminds us of our contrasting ability to live and breathe freely, to move and act of our own accord. The power of the notion of humans animating the inanimate world hinges on the assumption that nature and the automaton are symmetrical, but in reality the asymmetry between the object and the world cannot be overcome. Automaton and world are absolutely distinguishable. In other words, automata are not master simulators of nature; they do not exemplify the cohabitation of nature and artifice, a trait of so many of the objects that were strewn throughout princely collections. But, then, what are they? The automata discussed are doubly artificial; they are based on contemporary visual and material culture—liturgical vestments, paintings of the adoration of the Magi, manuscript marginalia, prints of theatrical and exotic state ceremonies, and mythological imagery. Thus the invention of automata in the early modern German-speaking world takes place within the history of art. To study them is not to open up the canon or to merely call out for these curiously ignored objects to be more closely studied by art historians. This book has attempted to show that early modern automata are *art historical.* They are above all self-conscious recontextualizations and reinterpretations of works of art.

To be sure, their moment in the sun was brief. Many were lost or melted down in the proceeding centuries. Perusing inventories of Kunstkammern that were compiled at the end of the seventeenth century, one is hard pressed to find automata among the plethora of porcelain, precious metalwork, natural specimens, and antiquities. But in the second half of the sixteenth century the clockwork automaton was greatly admired and sought after by high-ranking nobility in the Holy Roman Empire, and the extant objects and inventories of princely Kunstkammern attest to this. Their pervasiveness in early modern collections (recall that Holy Roman Emperor Rudolf II had more than twenty-five automata in the Kunstkammer in Prague) and their repeated mention in descriptions of those collections suggest that they were highly visible objects and that their display value was enhanced by the particular political or religious argument they enacted.

Yet there is more to early modern automata than their political or religious significance. The intended viewers of several automata described in the previous pages—Rudolf II, Christian I, and Murad III—were interpolated into the nefs, the Christmas Crib Automaton, and the *Sultan on Horseback* so that they would recognize themselves in the automata. Thus the locus of action was not just the object's movement but also the collapsing of the distance between the viewer and the object. The intended viewer was both the agent of the automaton's movement and a participant in the scene that the automaton set into motion. But the automata carefully engage a wider audience as well. We become a witness to the liturgical ceremony that transformed Charles V into a *persona gemina*, someone human by nature but divine by grace, on both nefs; the Christmas Crib Automaton permits us to behold not only the ceremonial act of the three kings giving gifts to the Christ child but more importantly the kings' literal conversion, their turning toward Christ. Whereas the nefs and the Christmas Crib Automaton were models of order and balance, the Verkehrte Welt Automaton is an object that revels in chaos and confusion. Confronted with the bizarre iconography and the repetitive, incoherent gestures of the preaching ape, we become the listeners of a silent sermon. With the *Sultan on Horseback*, we become front-row spectators to the sultan's procession through the streets of Istanbul. Finally, with the Diana Automaton, we are transformed into the agent of Actaeon's dismemberment.

Finally, predictability was crucial to the rhetoric of the early modern automaton. On one level, the objects were programmed to perform the same sequence of actions every time they were wound up and put on a surface. But on another level, automata were intended to be predictable because none of them set in motion anything ordinary: they animated precisely calculated choreographed ritualized events—coronations, the bestowal of gifts, preaching, and processions—that were meant to run like clockwork. The early modern automata examined in this book animated behavior and events that were to be understood as timeless but in reality were variable and contingent. But like ritual acts that always run the risk of running afoul, so too the automaton was vulnerable to malfunction. As detailed in chapter 5, defects were not uncommon if automata were exposed to the elements or neglected. Therefore, like the ceremonies they animated, automata had to be safeguarded against contamination. By being mechanized, the ritual is theoretically cleansed of any potential slippages or unexpected interruptions, and can be repeated ceaselessly. Perhaps the automaton originated from an impulse to display a perfect and controllable event. Perhaps automata were created to escape the unpredictable condition of the world, not to re-create it.

Notes

Chapter 1

1. The word "automat" appears for the first time in German in 1621 in Johann Geiger's *Horologium politicum*: "Automatum, das ist ein solches künstliches, selbstgehendes und schlagendes Uhrwerk." Geiger, *Horologium politicum*. For other early examples of the use of the term in German, see Mayr, *Authority, Liberty, and Automatic Machinery in Early Modern Europe*. The world "automaton" appears in English for the first time in 1611. *OED*, s.v. "automaton."
2. For a list of automata found in inventories of princely Kunstkammern in Prague, Innsbruck, Munich, and Dresden, see Keating, "The Machinations of German Court Culture," 238–46. On the fate of precious metalwork in early modern Germany, see Stielau, "The Weight of Plate in Early Modern Inventories and Secularization Lists."
3. To date, little work has been done by art historians on automata. Automata ancient, medieval, and early modern have long been the purview of historians of science and technology, as well as intellectual historians. The literature in this vein is vast. Recent work includes: Berryman, "Ancient Automata and Mechanical Explanation"; Silvia Berryman, "The Imitation of Life in Ancient Greek Philosophy," in *Genesis Redux*, ed. Riskin. For recent work on automata in the medieval period, see Truitt, *Medieval Robots*. Scholarship that addresses the early modern period, as of late, has been dominated by literary historians and critics: see Alexander Marr, "*Gentille curiosité*: Wonder-working and the Culture of Automata in the Late Renaissance," in *Curiosity and Wonder from the Renaissance to the Enlightenment*, ed. Evans and Marr, 150–72; LaGrandeur, *Androids and Intelligent Networks in Early Modern Literature and Culture*; Hyman, ed., *The Automaton in English Renaissance Literature*; Kang, *Sublime Dreams of Living Machines*. For work that deals with automata during the Enlightenment, see Voskuhl, *Androids in the Enlightenment*; Riskin, "The Defecating Duck; or, the Ambiguous Origins of Artificial Life"; Schaffer, "Deus et Machina"; Schaffer, "Enlightened Automata," in *The Sciences in Enlightened Europe*, ed. Clark, Golinski, and Schaffer, 126–65; and Daniel Schulthess, "Zur Infinitisierung der Automaten: Descartes und Leibniz," in *Androïden*, ed. Söring.
4. Baudrillard, *Simulacra and Simulation*, 93.
5. On this particular automaton and the political circumstances under which it was given, see Lietzmann, "Die Geschichte zweier Automaten."
6. On Drausch, see Lietzmann, *Valentin Drausch und Herzog Wilhelm V. von Bayern*.
7. Thomas Winkelbauer, "Ständefreiheit und Fürstenmacht: Länder und Untertanen des Hauses Habsburg im konfessionellen Zeitalter (1522–1699)," vol. 6 of *Österreichische Geschichte*, ed. Herwig.
8. Nonnus, *Dionysiaca*, 29.193.
9. Homer, *The Iliad*, 18.414–421.
10. On the myth of Daedalus and its formative influence on the Greeks' notion of art, see Morris, *Daidalos and the Origins of Greek Art*.
11. Aristotle, *De Anima*, 406b 18–19.
12. Tybjerg, "Wonder-Making and Philosophical Wonder in Hero of Alexandria."
13. There is debate around Ctesibius's dates and the

attribution of his work. See Drachmann, *Ktesibios, Philon, and Heron*; Drachmann, *The Mechanical Technology of Greek and Roman Antiquity*. For a more recent assessment of Ctesibius and other ancient mechanics, see Roby, *Technical Ekphrasis in Ancient Science*.

14. Drachmann, *Ktesibios, Philon, and Heron*, 68–72.
15. Murphy, "Heron of Alexandria's *On Automaton Making*." Murphy's article contains a translation of Hero's treatise. See also Boas, "Hero's Pneumatica." On the automaton with Hercules, see Berryman, "The Imitation of Life in Ancient Greek Philosophy," 35–45.
16. Al-Jazarī, *The Book of Knowledge of Ingenious Mechanical Devices*, 3–4.
17. Donald R. Hill, introduction to *The Book of Knowledge of Ingenious Mechanical Devices*, by al-Jazarī, 5–7.
18. Al-Jazarī, *The Book of Knowledge of Ingenious Mechanical Devices*, 17–18.
19. As quoted by Jill Burke in Burke, "Meaning and Crisis in the Early Sixteenth Century," 79.
20. Ibid., 80–81.
21. Vasari, *The Lives of the Artists*, 292–93.
22. Giovanni Paolo Lomazzo, "Libro dei Sogni" and "Trattato dell'arte de la pittura, scoltura e architettura, in Scritti sulle arti," in *Scritti sulle arti*, ed. Ciardi, 1:153, 2:96.
23. As quoted in Burke, "Meaning and Crisis in the Early Sixteenth Century," 80.
24. Weitzman, *Solomon*, chap. 6.
25. Brett, "The Automata in the Byzantine 'Throne of Solomon.'"
26. Liudprand, Cremona, *The Embassy to Constantinople and Other Writings*, 153.
27. Brett, "The Automata in the Byzantine 'Throne of Solomon,'" 478–79.
28. "The First Mechanical Clocks," in *Encyclopedia of Time*, ed. Macey, 127–33.
29. On the clock at the Strasbourg Cathedral, see Maurice, *Die deutsche Räderuhr*, 1:38; F. C. Haber, "The Cathedral Clock and the Cosmological Metaphor," in *The Study of Time II: Proceedings of the Second Conference of the International Society for the Study of Time, Lake Yamanaka, Japan*, ed. Fraser and Lawrence, 339–416; and F. C. Haber, "The Clock as Intellectual Artifact," in *The Clockwork Universe*, ed. Maurice and Mayr, 16–18.
30. Dasypodius, *Warhafftigte Ausslegung und Beschrybung des astronomischen Uhrwercks zu Strassbourg*.
31. Cipolla, *Clocks and Culture*, 42.
32. Ibid., 42.
33. Chapuis, Haraucourd, and Gélis, *Le monde des automates*, 1:167; Joseph Gallmayr, *Münchner Intelligenzblatt* (1779), 273. Gallmayr himself was a clockmaker and constructer of automata. In 1744 he built a clock made from wood in which the planets and figures moved. He is also reported to have designed and manufactured "dancing automata," celestial globes, and canaries that sprang and sang. See "Gallmayr, Joseph," in *Allgemeine deutsche Biographie*, 8:332.
34. Anthony Grafton, "The Devil as Automaton: Giovanni Fontana and the Meanings of a Fifteenth-Century Machine," in *Genesis Redux*, ed. Riskin, 46–62.
35. Marr, "*Gentille curiosité*."
36. On Turriano and the automaton of the monk housed in the Smithsonian, see Elizabeth King, "Perpetual Devotion: A Sixteenth-Century Machine," in *Genesis Redux*, ed. Riskin, 263–92.
37. García-Diego, *Juanelo Turriano*, 101.
38. "Ein ander werckh vom C. Marggraven mit veyelbraunem samet überzogen, is ein waxin bild, schlegt uff der cittern, mit dem kopf und hand sich bewegendt, man zeuchts auf wie ein uhrwerckh, schlect ein fantasia." Bauer and Haupt, eds., "Das Kunstkammerinventar Kaiser Rudolfs II., 1607–1611," f. 338 2145. On Margraf, see Bertele and Neumann, "Der kaiserliche Kammeruhrmacher Christoph Margraf und die Erfindung der Kugellaufuhr."
39. Cipolla, *Clocks and Culture*, 50.
40. For an excellent account of the development of clockmaking and the clockmakers' guild in Augsburg, along with transcriptions and translations of guild ordinances, see Eva Groiss, "The Augsburg Clockmaker's Craft," in *The Clockwork Universe*, ed. Maurice and Mayr, 57–86.
41. Cipolla, *Clocks and Culture*, 62.
42. Chapuis, Haraucourd, and Gélis, *Le Monde des automates*, 165–66.
43. On the Golden Bull, see Bryce, *The Holy Roman Empire*, 201–3; and Heer, *The Holy Roman Empire*, 116–17.
44. Ramus, *Scholarum mathematicarum libri unus et triginta*, 62; and Dee, *The Mathematicall Praeface to the Elements of Geometrie of Euclid of Megara* (1570).
45. "Ein uhr oder rederwerckh, ist ein pfaw, geht und wendt sich ringsumb, shreitt und macht ein wannen mit seinem schwaiff von rechten federn, steht

auff der tafel der kc." Bauer and Haupt, eds., "Das Kunstkammerinventar Kaiser Rudolfs II., 1607–1611," f. 338 no. 2142.

46. On the *Habsburger Pfau*, see Beate Kellner, "Formen des Kulturtransfer am Hof Kaiser Maximilian I: Muster genealogischer Herrschaftlegitimation," in *Kulturtransfer am Fürstenhof*, ed. Müller, 52–130, esp. 60.
47. Carmen Brown, "Bestiary Lessons on Pride and Lust," in *The Mark of the Beast*, ed. Hassig, 53–70, esp. 61–62.
48. *Deutsches Wörterbuch von Jacob Grimm und Wilhelm Grimm*, s.v. "Kür, kur, chur."
49. Bryce, *The Holy Roman Empire*, 274–86.
50. "The Religious Peace of Augsburg," trans. Emil Reich, in Reich, *Select Documents Illustrating Mediaeval and Modern History*, 230.
51. Gévay, *Urkunden und Actenstücke zur Geschichte der Verhältnisse zwischen Österreich, Ungern und der Pforte im XVI. und XVII. Jahrhunderte aus Archiven und Bibliotheken*, 3–23. For recent work, see Fichtner, *Terror and Tolerance*, and Johnson, *Cultural Hierarchy in Sixteenth-Century Europe*.
52. Malekandathil, *The Germans, the Portuguese, and India*.
53. Bauer and Haupt, eds., "Das Kunstkammerinventar Kaiser Rudolfs II."
54. For descriptions of the Kunstkammer in Dresden, see Beutel, *Chur-fürstlicher sächsischer stets grünender hoher Cedern=Wald auf dem grünen Rauten=Grunde, oder . . .* (Dresden, 1671), n2, and Hainhofer, "Des Augsburger Patriciers Philipp Hainhofer Reisen nach Innsbruck und Dresden," 176.
55. On the contents of the Gewölbe, see Seelig, "Die münchner Kunstkammer," 104–5. Johannes Fickler recorded the object in the 1598 inventory of the Kunstkammer: "Auf einem uberlengeten stöckl, von Hebeno ein grien geschmelzte berg, darauf ein viereckhet gulden gestell in die vierung mit rundafelin versezt, auf welchem sizt ein Aff von gold, geschmelzt, mit einer blawen Kappen, und ein Paketen in der handt, vor im ein gulden Pulpit, daruaf ein gesangbuech, der wirt von einem urwerckh bewegt, das der Aff mit der Paketen die Mensure schlegt. Neben dem Affen ligen ain Hirsch, ain stuckh wildt von goldt, waiss geschmelzt, ain gulden Rech. Oberst neben dem berg ain anderer geschmeltzer baum. Auf der seitens des stöckhels würt ein Deckhel fürgeschoben, darunder ein täfel in dem ein wald, darinen hirschen und Rech, von minature gemalt, in mitten am fürschub ain Porten an deren baiderseits ain Pyramis mit rubin am fuess versezt, oben auf geschmelzt, am Spiz ein Perlin hinder iedem ein Pyramis ain gulden turlen, auf der objerseitten zwai schubladen, so mit gulden geschmelzten hirsköpfl heraussgezogen werden. In dem grossern ligen ain guldene Reichpfeiffen, auch in dem clainern schublädl, das ein falsch, an idern ein geschmelzt hirshköpfl. Das fuetral hiezu ist inwendig mit weissem Attlass gefüertet, von aussen mit Plawen sammet uberzogen." Diemer, Bujot, and Diemer, eds., *Johannes Baptist Fickler das Inventar der Münchner herzoglichen Kunstkammer von 1598*, no. 3390. See also Diemer, Diemer, and Seelig, eds., *Die münchner Kunstkammer*, vol. 2, part 2, 1057–58; Hoyer and Stierhof, eds., *Schatzkammer der Residenz München*, 42.
56. Bredekamp, *The Lure of Antiquity and the Cult of the Machine*.
57. Todd Andrew Borlick, "'More than Art': Clock Automata, the Extemporizing Actor, and the Brazen Head in *Friar Bacon and Friar Bungay*," in *The Automaton in English Renaissance Literature*, ed. Hyman, 129–44.
58. Wendy Beth Hyman, "Mathematical Experiments of Long Silver Pipes: The Early Modern Figure of the Mechanical Bird," in *The Automaton in English Renaissance Literature*, ed. Hyman, 145–62.
59. Kang, *Sublime Dreams of Living Machines*, 55–102.
60. Riskin, "Machines in the Garden."
61. Marr, "*Gentille curiosité*."

Chapter 2

1. Menzhausen, "Kürfurst Augusts Kunstkammer"; Syndram, Minning, and Vötsch, eds., *Die Inventare der kurfürstlich-sächsischen Kunstkammer in Dresden: 1587*.
2. "Vorguldt kunstreich Schiff oder nave mit einer virtel und stunden schlagenden Uhr, welch alle 24 Stunden muss aufgetzogen werden, oben mit dreyen mastbaumen, uf welchen die Bussknechte im Mastkörben umbgehen, und die Viertel und Stunden auf den glöcklein mit hammern schlagen. Inwendigk die Rom. Key. Mayt. auf den Keyserlichen stul sitzendt, und vor deselben die sieben Churfürsten und Herolden mit erzeigung ihrer Reverenz zu endtfalunge der lehen umgehendt, Dessgleichen zehen Trommeter und ein Heerpauker, die da Wechselweise

zu tisch blasen, auch ein Drommelschleger und drey Trabandten sambt 16 kleinen stucklein, deren man II. Laden Khan, und von sich selbsten abgehen, darbey ein Futter, stehet auf einer grünen langen tafel mit tuch behengt." Hauptstaatarchiv Dresden (HStA) Inventare Nr. 1, fol. 254r.

The 1595 inventory of the collection lists the following: "Kunstlich mössen und vorguldt schieff oder Nave, mit einer virtel und stunden schlagen: den Uhr, welch alle 24. Stunden muss ufgezogen werden, oben mit drey mastbeument, auf welchen die Büssleuchte im Mast-Korben umbgehen, und die viertel und stunden uf den Glöcklein mit Hammern schlagen, Inwendigk die Rom. Key. Mayt: auf dem Keyserlichen stull sitzend und vor derselben die siben Churfürsten und Herolden mit erzeigung ihrer Reürentz zu entfahlunge der lehen umbgehendt, Dessgleichen 10 Trommetter und ein Heer: Bauker, auch ein Trommelschlager und sieben trabanten, samt 16. Kleinen stuckhen so man alle lossbrennen und abschliessen kann, und von sich selbst loss gehen, Darzu ein futter gehörigk." HStA Inventare Nr. 2, fol. 121r.

"Ain Schiff, so auf ainer tafel etliche bootsleuth darinnen forttreiben, wan es still stehet, aine thür sich aufthut, siben Churfursten des Reichs heraussgehen, Reverenz für Kaÿ. Maÿ bezeugen, Kaÿ. Maÿ. mit dem scepter und haupt gleichsam die lehen gibt, etlich trommeter wechselweisse blasen, ain heerpaucker auf den Kesselbauken schlegt in dem Mastbaum die Schlaguhren, und sonsten anderen bewegung vil zu sehen, vom Johann Schlothaimer zu Augsburg gemacht ist worden." This is how Philipp Hainhofer described the ship on his visit to the Dresden Kunstkammer in 1629. See Hainhofer, "Des Augsburger Patriciers Philipp Hainhofer Reisen nach Innsbruck und Dresden," 168.

3. "Nous allasmes voir en la maison d'un particulier, un horologue admirablement beau, auquel on travaille il I a desja seize ans. . . . Le maistre ouvrier inventeur d'iciuy, nommé Georg Roll, mourut il I a trois ans [en 1592], et i a maintenant un autre maistre qui le parachere. La boite de Cest horlogue est de beau bois d'ébaine, et est environ haute comme un picque, et I au bout d'icelle un coq. Que nous veismes chanter lors que l'heure sonnoit, i ayant au mesme instant un viellard fait de bronze, qui tourne un horologue. . . . Entre plusieurs figures d'hommes qui jouent leur personage en sait ouvrage, par le moyen des rouses et des resorts, on i voit l'Empereur en une chaire qui baisse la teste, devant les sept electeurs, qui luy vienent faire une grand reverance." Esprinchard, *Vie de Jacques Esprinchard, Rochelais, et journal des ses voyages au XVIe siècle*, 147–48.
4. "Ain hoches schönes alts gschirr, auf dem luckh siczt ain Kaiser in seinem tron under aim gwelb auf seiln, darunder ain uhrwerch, unden herumb die siben curfürsten, so da, wann uhrwerck gericht ist, im cürcl umbgeen und sich vor dem Kaiser naigen, kombt her von Cardinal Otto von Augsburg, wigt 39 marckh 6 lot." "Inventar des Nachlass Erzherzog Ferdinand II. In Ruhelust, Innsbruck und Ambras vom 30 Mai 1596," *Jahrbuch der Kunsthistorischen Sammlungen des Allerhöchsten Kaiserhauses* 7 (1888): xci–ccxxvi.
5. Pope Leo X (1475–1521) bestowed upon Charles the privilege of being crowned both king of the Romans (emperor-elect) and emperor within the same liturgical ceremony. Leo X also allowed the coronation to take place in his absence. This was the first and last time the pope would grant such a privilege. See Brandi, *Kaiser Karl V*, 100, and Fernández Álvarez, *Charles V*, 38.
6. Haspels, ed., *Royal Music Machines*, 197–99; Julia Fritsch, "Histoire, destinée, signification"; J. H. Leopold, "La Construction des nefs de Schlottheim"; and Eric Rieth, "L'horloge automate dite *Nef de Charles Quint* au regard de l'architecture navale du XVIe siècle," all in *Ces curieux navires: Trois automates de la Renaissance*, ed. Conihout, 1–34, 61–74, 99–111; Bobinger, "Der augsburger Uhrmacher Hans Schlothaim"; Maurice, *Die deutsche Räderuhr*, 2:229–30; Streng, "Augsburger Meister der Schmiedgasse um 1600." An eighteenth-century source claims that Schlottheim also made a similar automaton for Sultan Süleyman the Magnificent, but this has never been corroborated. See Stetten, *Kunstgewerbe und Handwerksgeschichte der Reichsstadt Augsburg*, 184.
7. Fritsch, "Histoire, destinée, signification."
8. Yates, *Astraea*, 1.
9. As quoted in Oman, *Medieval Silver Nefs*, 1.
10. "Item, la grant salière d'or, en façon d'une nef, que la ville de Paris donna au Roy, et est pareille à la grant nef don't cy-dessus est faicte mencion; pesant quinze marcs six onces d'or," and "Item, la grant nef du Roy, que la ville Paris luy donna, toute plaine; pesant VI xx V marcs d'or." Labarte, *Inventaire du mobilier de Charles V, roi de France*, 64, 77.

11. Ibid., 28.
12. Hoberg, *Die Inventare des päpstlichen Schatzes in Avignon*, 411.
13. Lightbown, *French Silver*, cat. no. 19, 28–34.
14. Oman, *Medieval Silver Nefs*, 13.
15. The literature on this miniature is vast. See Buettner, "Past Presents," 612, and Jean Pierre van Rijen, "Precious Metalwork in Gold Leaf: Everyday Lustre at the Court of Jean de Berry as Depicted by the Limbourg Brothers," in *The Limbourg Brothers*, ed. Dückers, 165–75.
16. Bobinger, "Der augsburger Uhrmacher Hans Schlothaim," 8.
17. Groiss, "The Augsburg Clockmaker's Craft," 57–58.
18. Little is known about Jeremias Metzger; one of his pieces survives in the Victoria and Albert Museum in London. It was signed by Metzger and has been dated to 1564. The marriage to Geiger presumably afforded Schlottheim social and occupational advancement, as the tools of a clockmaker and a locksmith in the late sixteenth century were by and large the same. In the sixteenth century the clockmakers were subsumed under the general guild of smiths, which included painters, saddlers, and goldsmiths. Archival documents often mention that locksmiths and clockmakers had a double qualification. Ibid., 60.
19. Maurice, *Die deutsche Räderuhr*, 120.
20. Streng, "Augsburger Meister der Schmiedgasse um 1600," 247–87.
21. Engelmann, "Das Krippenwerk des Augsburgers Hans Schlottheim"; Streng, "Augsberger Meister der Schmiedgasse um 1600," 273; and Maurice, *Die deutsche Rädruhr*, 125.
22. Fritsch, "Histoire, destinée, signification," 19.
23. Evans, *Rudolf II and His World*, 12. Soon after learning of his election to emperor, Charles traveled to England and then to Brussels, where he planned his coronation in Aachen. While in Brussels he was awarded one million florins for his coronation ceremony by the Estates General of the Netherlands. Reports out of Aachen claimed that the city was in the midst of a plague outbreak. Charles's advisors exhaustively strove to persuade him to have the coronation at a different site, such as Regensburg. The advisor's admonishments were ignored by Charles, who insisted on being crowned in Aachen regardless of how long he was required to delay his anointing. This decision, which was made during the most critical time to establish his dominance in Europe and position on the throne, clearly illustrates Charles V's reluctance to relinquish his reign's symbolic tie to the medieval empire. Charles was never indifferent to the location of his coronation. On the contrary, he displayed a keen sense of history, an awareness that heightened the profundity of his coronation. See Brandi, *Kaiser Karl V*, 100, and Fernández Álvarez, *Charles V*, 38.
24. Streng, "Augsburger Meister der Schmiedgasse um 1600," 153.
25. We do know that Rudolf commissioned objects while he was in Augsburg. A document dated December 31, 1582, from the imperial Hofzahlamt states: "Dem Elias Huetter, des Kurfursten von Sachsen Diener, welcher sich zur zeit des letzten Reichstages zu Augsburg dem Kaiser Rudolf II. zur Anfertigung einiger Wasserkunstwerke erboten hatte und von demselben im August mit dem Auftrage nach Wien geschickt worden war, die wasser kunstkwerke in der Gattermühle, zu Ebersdorf und an anderen Orten, besonders aber im Kaiserlichen Fasangarten zu giessen, nach Vollendung dieser Arbeit aber, um des Kaisers Gutachten abzuwarten, bis ende December 1582 zurückgehalten worden war und mit seinem Diener ausser den Augsburg erhalten 30 Gulden für Kost Wohnung und Arbeits Materialien 329 Gulden 20 Kreuzer augelegt hatte werden dieselben über Kaiserlichen befehl ersetzt." *Jahrbuch der Kunsthistorischen Sammlungen des Allerhöchsten Kaiserhauses* 7 (1888): clxix, no. 5415. To my knowledge we have no record of any object commissioned from Schlottheim during 1582.
26. During the early modern period this decorative motif was used in a wide range of settings, such as ceramics, furniture, textiles, armor, and frescoes. It was first employed north of the Alps by printmakers such as Israhel van Meckenem (ca. 1440–1503) and Martin Schöngauer (1450–1491). Remarkably, in an early printed portrait of Emperor Charles V by Daniel Hopfer (1470–1536), the emperor's bust is surrounded by the motif. Additionally, Hopfer engraved a map in which the entire geographical area of Burgundy, the first region Charles V inherited, is covered with *Laubwerk*. Thus, a tradition of surrounding the emperor with this decorative motif was established before Schlottheim manufactured the nefs and may have been a catalyst for the inclu-

sion of the motif on these objects. See Gruber and Pons, *L'Art décoratif en Europe*, 115–28.

27. The rigging of each of the surviving nefs is also fairly complex. They are rigged with three masts—the mizzen, main, and fore masts—at full sail. All the masts could be turned around in an attempt to simulate the wind blowing in the proper direction of movement. A shroud (a set of ropes that lead from the head of the mast and relieve lateral strain on the mast) extends from each mast. Three crows' nests—each containing a figure who strikes bells while the automata are in motion—are situated above each yard and sail, and crowned with trucks, spindles, and vanes on both works.

28. This insignie could only be used by the sovereign after his *conformatio* coronation by the pope in Rome. When the sovereign was elected the German King of the Romans in Frankfurt and later crowned in Aachen, he was depicted with the single-headed eagle. A famous example of this distinction can be found in Hans Burgkmair's *Kaiser Maximilian in der Kapelle* (1515). In the woodcut the mitre crown of the emperor is displayed over the double-headed eagle, while the double-arched crown of the emperor-elect is rendered above a single-headed eagle. For a discussion of the two crowns and their use before and during Charles V's reign, see Rosenthal, "Die 'Reichskrone,' die 'Wiener Krone,' und die 'Krone Karls des Grossen' um 1520"; Rosenthal, "Plus ultra, non plus ultra, and the Columnar Device of Emperor Charles V"; and Rosenthal, "The Invention of the Columnar Device of Emperor Charles V at the Court of Burgundy in Flanders in 1516."

29. Deshman, "Otto III and the Warmund Sacramentary," 8. The depiction of the emperor enthroned and in majesty began in the Holy Roman Empire during the Carolingian period. It is believed that the tradition was initiated by either Lothar I (795–855) in *The Gospel of Lothar I*, in the Bibliothèque nationale de France, Paris (ca. 849–51, Ms. Lat. 266, fol. 1v), or by Charles the Bald (823–877) in *The Psalter of Charles the Bald*, also in the Bibliothèque nationale de France (ca. 850, Ms. Lat. 1152, fol. 4v). The Ottonian Emperors appropriated this tradition and were regularly represented enthroned, not only in manuscripts such as the Aachen Gospels (ca. 975, Aachen, Aachner Dom Schatzkammer), but also on secular seals. "Otto I is the first ruler in Western sigillographic history to be depicted on his seal enthroned, crowned, holding a lily scepter and the Christological imperial orb surmounted by a cross." Bedos, "The King Enthroned," 60.

30. This notion is crystallized in the Norman Anonymous's eleventh-century *De consecratione pontificum et regum*: "We thus have to recognize [in the King] a *twin person*, one descending from nature, the other from grace. . . . One through which, by the condition of nature, he conformed with other men: another through which by the eminence of [his] deification and by the power of the sacrament [of consecration], he excelled all others. Concerning one personality, he was, by nature, an individual man: concerning his other personality, he was, by grace, a *Christus*, that is a God-man." As quoted in Kantorowicz, *The King's Two Bodies*, 46.

31. "In his reorganization of imperial government, Constantine introduced a high council of advisors that he called the 'consistorium,' literally 'the standing committee,' because when they convened only the emperor sat." Mathews, *The Clash of Gods*, 109.

32. Adelson, "The Holy Lance and the Hereditary German Monarchy," 181.

33. This curious addition of an elector has two possible explanations. First, the eighth elector could have been included during the time the automaton was crafted to compensate for a gap in the procession. Second, the eighth figure could have been affixed sometime after 1648, when an eighth electoral position was given to the duke of Bavaria, Maximilian I (1573–1651).

34. Hans Weiditz also published two earlier portraits of Charles, when he was King Charles I of Spain; the first dates to 1518 and the second to 1519.

35. "In 518 the consul Magnus sat on *a sella curulis* whose legs were lions' legs with lions' heads above them; the end panels of the front rail were decorated with personifications of abundance, and figures of victory danced on top of the perpendicular rails." Mathews, *The Clash of Gods*, 105.

36. For an excellent example of the Old Testament leonine throne that was incorporated into a coronation ceremony, see *Saul Anointed by Samuel* (ca. 1250), a detail from an illuminated manuscript from the Pierpont Morgan Library, New York (M. 638, fol. 23v).

37. On the Habsburg enterprise to manufacture their descent from Old Testament priest-kings, see Tanner, *The Last Descendant of Aeneas*, 80–83;

Wood, *Forgery, Replica, Fiction*; and Silver, *Marketing Maximilian*.

38. Kantorowicz, *The King's Two Bodies*, 126–27.
39. Kołakowski, *God Owes Us Nothing*, 34.
40. For more on the ramifications of this blur, see Agamben, *The Kingdom and the Glory*, 10–14.
41. Malebranche, *Treatise on Nature and Grace*, 195.
42. Kantorowicz, *The King's Two Bodies*, 47.
43. The pluviale, along with its counterparts—the *Marienmantel* and the *Johannesmantel* (both ca. 1525, Vienna, Weltliche Schatzkammer)—has been dated to the second quarter of the sixteenth century. See Eisler, "Two Early Franco-Flemish Embroideries," 578.
44. Certeau, *The Mystic Fable*, 1:81.
45. Augustine, "Enarrationes in Psalmos" (XXIII, 2, 183), in *Patrologiae cursus completus: Series latina*, ed. Migne, vol. 36.
46. Möbius, "Navis Ecclesiae," 15.
47. On the engraving, see Hind, *Early Italian Engraving*, 1:251–54, nos. EIII 6–7 and IV, plates 399–400; Levenson, Oberhuber, and Sheehan, *Early Italian Engravings*, 162–64; and Kemp, "Navis Ecclesiae."
48. Levenson, Oberhuber, and Sheehan, *Early Italian Engravings*, 162.
49. On the Tiburtine Sibyl, see Holdenried, *The Sibyl and Her Scribes*, esp. xvii–xxvi, and McGinn, *Visions of the End*, 43–44.
50. Eusebius, *Commentary on the Psalms*, in *Patrologiae cursus completus: Series graeca*, ed. Migne, vol. 23, 71.
51. Agamben, *The Kingdom and the Glory*, 4–9.
52. In addition to two suspected failed attempts at poisoning Rudolf II's father, Maximilian II, Philip II sought an imperial title to rival that of his Austrian relatives. In January 1563 rumors were flying around Europe that he was preparing to declare himself emperor of the Indies, and a similar rumor circulated just four months later that he had proclaimed himself "King of the Indies and of the New World." Twenty years later a French ambassador in Venice wrote to the French king Henri II to inform him that Antoine Perrenot de Granvelle (1517–1586), one of Philip's leading jurists and ministers, had visited the pope in Rome to discuss an imperial title for Philip II. "I have learned from these Lords that Cardinal Granvelle is coming to Rome in September this year to have the title of emperor conferred upon his master." As quoted in Braudel, *The Mediterranean and the Mediterranean World in the Age of Philip II*, 675.
53. *Jahrbuch der Kunsthistorischen Sammlungen des Allerhöchsten Kaiserhauses* 7 (1888): clxix, no. 5370.
54. The Rechnungen from the Hofzahlamt in Prague recorded that on May 19, 1581, Rudolf II sent the money to Spain for the portraits. "Der Hofzahlmeister entrichtet an Herrn Hans Khevenhüller, Kaisers Rudolf II. Rath und Gesandeten des Königs von Spanien, 117 Gulden 20 Kreuzer für ein Gemälde des Kaisers Karl, welches für den Kaiser gemacht wurden und noch aus Spanien erwartet wird." *Jahrbuch der Kunsthistorischen Sammlungen des Allerhöchsten Kaiserhauses* 7 (1888): clvvv, no. 5382.
55. The location of the busts is discussed in Lars Olof Larsson, "Portraits of Emperor Rudolf II," in *Rudolf II and Prague: The Court and the City*, ed. Fučiková, 127.
56. Fernández Álvarez, *Charles V*, 34.
57. My brief description has been taken from Karl Brandi. See Brandi, *Kaiser Karl V*, 105–9.
58. For an overview of the roles coronations play in societies and a historiographic discussion of the topic, see János M. Bak, "Introduction: Coronation Studies—Past, Present and Future," in *Coronations*, ed. Bak, 1–17. As one might imagine, the scholarship on this ritual event is immense. The author who pioneered this field of study, Percy E. Schramm, was primarily interested in divulging the mechanisms that promoted the symbolic nature of kingship in the medieval period. Numerous students of Schramm have published on coronations. Their collective body of work is referred to as the Göttingen School. Ernst Kantorowicz's masterly *The King's Two Bodies* was written in response to Schramm's work. See Schramm, *Herrschaftszeichen und Staatssymbolik*, vol. 2, and Kantorowicz, *The King's Two Bodies*.
59. Shortly after Charles's election, his head jurist, Mercurino Gattinara (1465–1530), wrote to him and stated: "Sire, now that God in His prodigious grace has elevated Your Majesty above all Kings and Princes of Christendom, to a pinnacle of power occupied before by none except your mighty predecessor Charlemagne, you are on the road toward Universal Monarchy and on the point of uniting Christendom under a single shepherd." As quoted by Fernand Braudel in *The Mediterranean and the Mediterranean World in Age of Philip II*, 674. Charles V's rule even prompted nervous papal authorities to concede that the young emperor had made the medieval notion of a universal Christian monarchy (*monarchia universalis*) a reality. On the repeated com-

parison made between Charles V and Charlemagne, see Bosbach, "Die politische Bedeutung Karls des Grossen für Karl V," 49–73. On the response of the papal authorities, see Kohler, *Karl V. 1500–1558*, 194.

60. Kaufmann's work on the Rudolfine court is extensive. The primary texts in which he lays out his thesis on the imperial theme at the Habsburg court are: *Variations on the Imperial Theme in the Age of Maximilian II and Rudolf II*; "Remarks on the Collections of Rudolf II"; *The School of Prague*; *The Mastery of Nature*; and *Arcimboldo*.
61. For the dating of *Vertumnus* and its relationship to the poems that accompanied it, see Kaufmann, *Arcimboldo*, 61. It should be noted that Kaufmann formulated his "variation on an imperial theme" in response to Frances Yates's work on themes of empire in the sixteenth century. See Yates, *Astraea*.
62. On portraiture *ex se*, see Koerner, *The Moment of Self-Portraiture in German Renaissance Art*, 9, 175.
63. Kaufmann, *Arcimboldo*, 93.
64. Barthes, *Arcimboldo*, 15–16.
65. Kaufmann, *Arcimboldo*, 93.
66. Evans, *Rudolf II and His World*, 102.

Chapter 3

1. The 1610 inventory lists the following: "die Geburt Christi von Kupfer gemacht, versilbert und vergoldet und mit allerlei Bildwerk gezieret . . . ist Churfürst Christian, hochlöblicher Gedächtnis von derselben geliebten Gemahlin zum Heiligen Christ verehret worden anno 1589." HStA Inventar Nr. 3, fol. 363v. The piece was also recorded by Tobias Beutel in 1611: "Der funftlich beweglichen Sachen und Uhrwercke werden zum wenigsten hundert stuck gezehlet / die vorneumsten sind: Das grosse Astronomische Uhrwerck / so Chur=fürst Augsto (hochstseeligsten Andectens) 1600.Rthlr.gecostet; Ein Uhrwerk von der Geburt Christi; Zwen in Form wie Schiffe / als Papegonen / ein als ein Pfau / Ouctguct / Lamb / und andere thiere." Beutel, *Chur-Fürstlicher sächsischer stets grünender hoher Cerdern=Wald auf dem grünen Rauten-Grunde*, n. 2. Philipp Hainhofer also described the object, which he saw in 1629 during his visit to the Dresden Kunstkammer: "Ain schönes vhrwerck, darinnen die geburt Christj samt der raÿse der weysen aus Morenland, wie auch der engel herabfahrung vom himmel, ochs und esels sprung, suchung und anbetung her hirten zu sehen, auch etlich Weihenact gesang durch ain pffeifenwerck zu hören, vom Schlotthammer." Hainhofer, "Des Augsburger Patriciers Philipp Hainhofer Reisen nach Innsbruck und Dresden," 176. The 1640 inventory lists the following: "Ein schon gross uhrwergk, darinnen die Geburth Christi wie auch der Hirten und Weisen auf dem Morgenland ihre Bildwergk zu finden, welch in ablauffung des Uhrwergks herumb gehen und sich voer dem Kindlein neigen, oben auch sich mit zweyen flügeln voneineander thut, die Engle gleichsamm wie vom Himmel herabgflihen undt durch ein Pfeifenwergk, 'Vom Himmel hoch da komm ich hehr' singenn: Entlich Joseph wiegt und Maria durch ein Pfeifwergk singet 'Joseph lieber Joseph mein.' Is alles von Kupffer und Messing vergüldet und versilbert, und ist Churfürst Christiano I zu Sacssen Hochseeligten gedechtnis und deroselben vielgeliebten Gehahl zum Heyligen Christ verehrt wordern. Anno. 1589." HStA Inventar Nr. 4, fol. 496r.
2. Former location: Mathematisch-Physikalischer Salon, Dresden (destroyed). Literature: Peter Plassmeyer, "Renaissance Musical Automata in the Art Collection of the Saxon Electors in Dresden," in *Royal Music Machines*, ed. Haspels, 49; Maurice, *Die deutsch Räderuhr*, 2:389; Protz, *Mechanische Musikinstrumente*, 36–38; Engelmann, "Das Krippenwerk des Augsburgers Hans Schlottheim."
3. Lietzmann, "Die Geschichte zweier Automaten."
4. Buettner, "Past Presents," 613.
5. On the New Year's festivals at late medieval and early modern courts, see Buettner, "Past Presents," 598–619; Casey, "Court Performance in Berlin of the Sixteenth Century"; Collins, ed., *Jewels and Plate of Queen Elizabeth I*, 101–8; Gregory Lubkin, "Christmas at the Court of Milan: 1466–1476," in *Florence and Milan*, ed. Smyth and Garfagnini, 2:257–70; Syndram and Scherner, eds., *Princely Splendor*, 11–12; and Kretzenbacher, *Frühbarockes Weihnachtsspiel in Kärnten und Steiermark*, 7–32.
6. Syndram, *Renaissance and Baroque Treasury Art*, 38–39; Klaus Pechstein, "Der Goldschmied Wenzel Jamnitzer," and Eduard Isphording, "Wenzel Jamnitzer und sein Werk im Urteil der Nachwelt," in *Wenzel Jamnitzer und die nürnberger Goldschmiedekunst*, ed. Forssman, 57–70, 191–206. On the other Jamnitzer Prunkkassette in the Dresden Kunstkammer, see Forssman, "Die Prunkkassette mit der Allegorie der Philosophie von Wenzel Jamnitzer." Additionally, it is important to note that Sophie's stepmother,

Elizabeth Kurfürstin von Brandenburg, gave Christian I yet another Prunkkassette on Februrary 28, 1590, as a New Year's gift. However, this piece was not as costly an item as Jamnitzer's case. Even though the "Kestlein oder Nöheledtlein," as it was referred to in the 1595 inventory of the Dresden Kunstkammer, followed the style of Jamnitzer, the case Elizabeth gave to Christian incorporated linen and embroidered work in addition to costly silver castings. See Syndram and Scherner, eds., *Princely Splendor*, 288–91.

7. "A lover may freely accept from her beloved these things: a handkerchief, a hair band, a circlet of gold or silver, a brooch for the breast, a mirror, a belt, a purse, a lace for clothes, a comb, cuffs, gloves, a ring, a little box of scent, a portrait, toiletries, little vases, trays, a standard as a keepsake of the lover, and to speak more generally, a lady can accept from her love whatever small gift may be useful in the care of her person, or may look charming, or may remind her of her lover, providing, however that in accepting the gift it is clear that she is acting quite without avarice." Andreas Capellanus, as quoted in Camille, *The Medieval Art of Love*, 51. For an excellent catalog of many of the courtly cases, see Kohlhaussen, *Minnekästchen im Mittelalter.*

8. "A model of comportment, for example, might be a book of manners that defines proper behavior in a variety of difficult social situations. A model might be a 'how to' book that explains how to solve a calculus problem, repair a car, or bake a cake. To follow such a model means to think certain thoughts or perform certain actions in accord with the rules the model presents. Second, a model might consist of a miniature of something. . . . The miniature model presents a simplified and closed example that parallels the confusing complexity of the thing it models, but also a model anticipates the future in some way: it allows one to imagine creating something in its image. . . . Mirrors, on the other hand, present the world as it is understood to be. They have a declarative character. . . . Unlike a model they do not offer an alternative for the future constitution of society." Muir, *Ritual in Early Modern Europe*, 5.

9. Karlheinz Blaschke, "Religion und Politik in Kursachsen, 1586–1591," in *Die reformierte Konfessionalisierung in Deutschland*, ed. Schilling, 79–97.

10. Katrin Keller has pointed out, "Dieses vom neuen, Lutherischen Eheideal und Rollenverstandis ganz offensichtlich beinflusste Bild setzen Anna und August als Ausdruck ihrer Frömmigkeit wie als deizidiert evangelisches Leitbild bewusst ein, um sich als Herrscherpaar des neun reformatorischen Zeitalters darzustellen. Zu diesem Leitbild gehörte freilch auch die Einigkeit in allen Gragen des Wirken über Familie und Haushalt hinaus und selbstverstandlich die Zurückhaltung der hinsichtlich diesbezüglicher Aktivitäten." Katrin Keller, "Kurfürstin Anna von Sachsen (1532–1585): Von Möglichkeiten und Grenzen einer 'Landesmutter,'" in *Das Frauenzimmer: Die Frau bei Hofe in Spätmittelalter und Früher Neuzeit*, Hirschbiegel and Paravicini, eds., 266.

11. Watanabe-O'Kelly, *Court Culture in Dresden*, 15.

12. The marriage alliance between the elector of Saxony and the elector of Brandenburg was already apparent in 1581, when August I took Christian to Cölln an der Spree to participate in the festivities surrounding the baptism of the elector of Brandenburg's son, also named Christian. A printed account of this visit was written by Philipp Agricola at SLUB Hist.Brand. 14, misc. 2.

13. Elector Johann Georg of Brandenburg did not tolerate Calvinism or any reformed ideas within his domain. "Schon auf eine dahin zielende Bitter der Stände gab er denselben 1572 die Zusicherung, dass die Lehre des göttlichen Wortes, wei sie durch Dr. Luther bei seinem Leben gelehrt, allein und ausschliesslich im Lande gelten und keine andere Lehrmeinung oder Ceremoni geduldet werden solle." L. H. Hirsch, "Johann Georg, Kf. v. Brandenburg," in *Allgemeine deutsche Biographie*, 14:167.

14. The exportation of Sophic was for the Brandenburg house the most politically advantageous marriage of her generation. Her brother and future elector Joachim Friedrich I was married in 1570 to Katharine von Brandenburg-Küstrin. Sophie's sister Erdmuthe was married in 1577 to Herzog Joachim Friedrich von Pommern. Another sister, Anna Marie, was married to Barnim XII, yet another Herzog from Pommerania, in 1581. Her brother Christian was married to Marie, the daughter of Herzog Albrecht Friedrich von Preussen, in 1604. A third sister, Magdalene, was married to Ludwig, Landgraff von Hessen-Darmstadt, in 1598. And a third brother, Joachim Ernst, was married to Sophie von Solms-Laubach in 1612. Sophie's younger sister Agnes was married in 1604 to Philipp Julius,

the Herzog von Pommern. Elisabeth Sophie was married to Janus I von Raidziell. And finally, her youngest surviving sister, Dorothea Sibylle, married Johann Christian Herzog von Liegnitz. Needless to say, all these marriages that Johann Georg arranged for his numerous children—twenty-two in all, of whom ten survived to adulthood—were to courts that did not have the political strength of the Saxon court. See Isenburg, *Stammtafeln zur Geschichte der europäischen Staaten*, vol. 1, *Die deutschen Staaten*, 62, 65–66.

15. The medal has been published in Holzhausen and Watzdorf, eds., "Brandenburgisch-Sächsische Wachsplastik des XVI. Jahrhunderts: Studien aus den Kunstkammern in Berlin und Dresden," 248–49.
16. HStA Geheimes Archiv 8017/12, docs. 2, 3.
17. For an excellent discussion of the reception of the Magi during the early modern period, see Trexler, *The Journey of the Magi*.
18. In the beginning of his reign, August also openly favored Philippists, followers of Melancthon, over orthodox Lutherans. Ernst Koch, "Der kursächsische Philippismus und seine Krise in den 1560er und 1570er Jahren," in *Die reformierte Konfessionalisierung in Deutschland*, ed. Schilling, 60–77.
19. Nicollier-De Weck, *Hubert Languet (1518–1581)*, 181–89.
20. Watanabe-O'Kelly, *Court Culture in Dresden*, 15.
21. Christian I's official title was "Zivilgouverneur." See Chrisa Schille, "Christian I," in *Neue deutsche Biographie*, vol. 3, 230.
22. Bäumel, "'Cave Calvinae—D.N.K.'"
23. Gotthard, "'Politice seint wir bäpstisch'"; Watanabe-O'Kelly, *Court Culture in Dresden*, 23.
24. The base's Italianate architectural elements and their arrangement strongly resemble the portal of the Dresden Residenzkapelle, which dates to about 1555. The portal now stands in the Judenhof in Dresden and is the only surviving architectural feature from the chapel. It was most likely designed and partly executed by the Italian Johann Maria. The statues of the two Saint Johns, Saint Peter, and Moses that stand in the niches of the portal were carved by him. He may also have executed the large relief in the attic and the statue of Christ flanked by two angels (originally four) carrying the symbols of the passion. Hitchcock, *German Renaissance Architecture*, 103–4.
25. B. Wachinger, "Mönch von Salzburg," in *Die deutsche Literatur des Mittelalters*, ed. Ruh, Langosch, Stammler, and Stöllinger-Löser, 6:658–70.
26. Bacon, ed., *Dr. Martin Luther's deutsche geistliche Lieder*, xxix.
27. Kretzenbacher, *Frühbarockes Weihnachtsspiel in Kärnten und Steiermark*, 11.
28. For instance, New Year's greetings were given to the audience by a child whose name was Wilhelm von Lewen; they employed the low German dialect. "Vill glug Euer Gnaden wiederfhar / In anfang, zu dem Newen Jhar / Gott lass Euer Gnaden werden zu Theill / Da New geborne Kindelein / Dauon wir Jtz bringen herein / Ein Spill, Kurtz, schlecht gering, undt klein." As quoted in Casey, "Court Performance in Berlin of the Sixteenth Century," 57.
29. Lamb, *Josef lieber, Josef mein*, 2.
30. HStA Geheimes Archiv 8017/12, docs. 2, 3.
31. "Vorguldt kunstreich Schiff oder nave mit einer virtel und stunden schlagenden Uhr, welch alle 24. Stundent muss afgetzogen werden, oben mit dreyen mastbaumen, uf welchen die Bussknechte im Mastkörben umbgehen, und Viertrelund Stunden auf den glöcklein mit hammern schlagen. Inwedigk die Rom. Key. Mayt. aud den Keyserlichen stul stizendt, und vor deselben die sieben Churfürsten und Herolden mit erzeigung ihrer Reverenz zu entfalunge der lehen umgehendt, Dessgleichen zehen Trommeter und ein Heerpauker, die da Wechselweise zu tisch blasen, auch ein Drommelschleger und drey Trabandten sambt 16 kleinen stucklein, deren man II. Laden Khan, und von sich selbsten abgehen, darbey ein Futter, stehet auf einer grünen langen mit tuch behengt." 1587 Inventar, Blatt 254r.
32. This particular tankard in the Dresden Kunstkammer is based on an image from Schedel's *Weltchronik* of 1493 and a later woodcut by Hans Vogel. The emperor is enthroned in the center of the vessel and flanked on his right by the three ecclesiastical electors: Mainz, Cologne, and Trier. To the left of the emperor stand the four "secular" offices of electors: the king of Bohemia, the elector of Palatine, the elector of Saxony, and the elector of Brandenburg. Oftentimes this object and others similar to it are referred to as "glasses with the imperial eagle." Almost all these tankards were painted in the city of Kreibitz in Bohemia. This particular piece is the earliest of its type. It also has a lengthy inscription. Above the electors it reads: "Die Römische kayserliche Mayestät, sampt Den

Sieben Churfürsten: Inn Irer Kleidung Ampt und Sitz etc." Below the figures it reads: "Der ertzbischoff zu mentz bekande, ist Cantzler im Deutschen lande. Sonst der bischoff zu Cöln gleich / Cantzler durch gantz franckreich. Darnach der ertzbischoff zu trier, Is Cantzler in welscher refier. 1588 Also in all ihrem Ornat / Sitzt kayserliche Mayestat, Sampt den sieben Churfürsten gut wie den ein ider sitzen thut. Inn Churfürstlicher kleidung sein / Mit an Zeigung des ampts sein. Der König in Beheim der ist / Des reichs erzschenck zu aller frist. Hernach der pfaltzgraff bein rein Des heiligen reich truchsäss thut sein. Der Hertzog zu Sachsen geborn Ise des reichmarschalch auserkorn. Der Marggraff von Brandenburg gut / des reich ertzkämmer sein thut." See Haase and Karpinski, *Sächsisches Glas*, 295 cat. nos. 8 and 9.

33. Koerner, "The Epiphany of the Black Magus circa 1500," in *The Image of the Black in Western Art*, Bindman, Gates, and Dalton, eds., 11–12; Trexler, *The Journey of the Magi*, 93.
34. Queen Elizabeth famously gave out thousands of gifts, most of gilt plate, during the New Year's festival. She gave gifts to people she received gifts from and even to some she did not receive gifts from, such as the grooms of her Privy Chamber, the yeoman of the robes, and the Jewel House officers. See Collins, ed., *Jewels and Plate of Queen Elizabeth I*, 103.
35. Bächtold-Stäubli, ed., *Handwörterbuch des deutschen Aberglaubens*, 1027.
36. Trexler, *The Journey of the Magi*, 78–84.
37. Buettner, "Past Presents," 600; Collins, ed., *Jewels and Plate of Queen Elizabeth I*, 103–8; Lubkin, "Christmas at the Court of Milan," 258.
38. *Princely Magnificence*, 12.
39. Lubkin, "Christmas at the Court of Milan," 258.
40. Buettner, "Past Presents," 605.
41. Cahill, "The Treasury at Persepolis," 375.
42. Kaufman, "Luther's 'Scholastic Phase' Revisited," 282.
43. Martin Luther, "On Threefold Righteousness," in *The Works of Martin Luther*, vol. 1, trans. Henry Eyster Jacobs (Philadelphia: A. J. Holman, 1915), 89.
44. Martin Luther, "The Freedom of the Christian," in *The Works of Martin Luther*, vol. 2, trans. Jacobs, 393.
45. Ben-Amos, "Gifts and Favors," 223.
46. Martin Luther, "The Explanation of the Theses," in *The Works of Martin Luther*, vol. 3, trans. Jacobs, 19.
47. Ernst Ullman, "Die Wittenburger Unruhen, Andreas Bodenstein von Karlstadt und die Bilderstürme in Deutschland," in International Congress of the History of Art, *L'art et les révolutions*, 117–226.
48. Martin Luther, "Against the Heavenly Prophets in the Matter of Images and Sacraments," in *Theological Aesthetics*, ed. Thiessen, 134.
49. Ibid., 132.
50. Ibid., 133.
51. As quoted by Sergiusz Michalski in Michalski, *The Reformation and the Visual Arts*, 27.
52. "Dan wiewol uns Gottes wort / in der heiligen Schrifft. Gnugsam und reichlich lehret / was wir thun und lassen sollen . . . Dannoch bedörffen wir / umb unserer schwacheyt willen / auch gutter exempel / und fürbilde / der hellige[n] / und geist reichen leute / darin[n] wir eben dasselbig sehen / das uns Gottes wort lehret / von denen wir lernen / und denen wir nachfolgen solle[n] / damit wir unser leben unstafflich unnd Christlich füren." As quoted in Bridget Heal, "Sacred Image and Sacred Space in Lutheran Germany," in *Sacred Space in Early Modern Europe*, ed. Coster and Spicer, 53 n. 52 (my translation).
53. Gutfleisch and Menzhausen, introduction to "'How a Kunstkammer Should Be Formed,'" 3.
54. Ibid., 4.
55. Ibid., 10.
56. Ibid., 8.
57. Blaschke, "Religion und Politik in Kursachsen, 1586–1591," 95–96.
58. "Vor einen grossen Spiegell, welcher mit 1000 lot vergültten silber beschlagen. Auch mit des gantzen Romischen reichs wappen, und viehlen darien verstzten Bemischen steihnen getziert." Quoted in Syndram, *Schatzkunst der Rennaissance und des Barock*, 59 (my translation).
59. Hayward, *Virtuoso Goldsmiths and the Triumph of Mannerism, 1540–1620*, 253–55; Hayward, "The Mannerist Goldsmiths," 22–30; Schröder, *Dirich Utermarke ein hamburger Goldschmied der Renaissance*; Tipton, *Res publica bene ordinata*, 44–50.
60. The initials of the makers appear on two shields attached to the mirror. One of the shields is topped with blossoms and a sickle, which are flanked on each side by the initials LM. The second shield has a rampant lion flanked by the initials DV. Hayward, *Virtuoso Goldsmiths and the Triumph of Mannerism, 1540–1620*, 389.
61. Utermarke originally lived in Lüneburg, and it is here where he met Meyer. Utermarke was not

a member of the goldsmiths' guild in Lüneburg but instead belonged to the brewers' guild and worked illegally for Meyer on occasion. In 1592 Utermarke presented a piece to the goldsmiths' guild in Lüneburg, but it was rejected. After being turned away from the guild, Utermarke left the city and settled in Hamburg, where he worked as a *Freikünstler* (independent master). In 1595 Utermarke presented a piece to the goldsmiths' guild in Hamburg. This time he succeeded in being inducted into the guild, and he spent much of his time there making presentation cups for the city of Hamburg (seventy-five cups survive, along with twelve gilt spoons). Schröder, *Dirich Utermarke ein hamburger Goldschmied der Renaissance*, 100–102.

62. "1 Grosser von silber und vergüldter spiegell mit böhimischen steinen gezierrt, darahn die gantz Prophezíung Danielis von den vier Monarchien, auch des Römischen Reiches unnd der darein gehörigen Königreichen, lender unnd Provincien Wappen, ist von der Churfürstlichen Sächsischen Wittwe, von einen Lüneburger erkaufft worden, und is das glass darauf die Wappen amaliert zerschrichtt." HstA 1610 Inventar, fol. 167r.
63. Syndram and Scherner, eds., *Princely Splendor*, 60.
64. Tipton, *Res publica bene ordinata*, 45.
65. Hentschel, "Epitaphe in der alten dresdner Frauenkirche," 101–25. An engraving of the Schönberg epitaph was included in Johann Gottfried Michaelis's *Dressdnische Inscriptiones und Epitaphia*. The remnants of the sculpted epitaph are now in the Dresden Stadtmuseum.
66. Deri, *Das Rollwerk in der deutschen Ornamentik des sechzehnten und siebzehnten Jahrhunderts*, 4.
67. Daniel 2:39.
68. Thomas Rahn, "Geschichtsgedächtnis am Körper: Fürstliche Merk- und Mediationsbilder nach der Weltreiche-Prophetie des zweite Buches Daniel," in *Seelenmaschinen*, ed. Berns and Neuber, 521–61. It is also important to note that Luther was often rendered in the guise of Daniel. A painting from the second half of the sixteenth century, which hung in the Eisleben Luther-House, depicted Luther before the emperor at Worms. In the image Luther refused to worship an idol that looked surprisingly similar to the emperor himself. Scribner, "Incombustible Luther," 54.
69. "Es muss bleiben bis an Jüngsten Tag, wie schwach es simmer sey, denn Daniel leugnet nicht und bischer die Erfahrung auch beweiset hat, beide an Bebsten und Königen." Martin Luther, "Vorrede uber den Propheten Daniel," in *Biblia*, ed. Füssel, vol. 2 Ir.
70. The 1587 inventory of the Dresden Kunstkammer also records in the library an "Anatomia Statuae Danielis, in gestalt eines grossen gewapneten Mannes, mit den Bedeutungen der 4 Monarchien sambt dem Sachssischen Stamme, in einmem eingefassten Rahmen mit Marmolfarbe angestrichen, hat Laurentius Faust pastor Schirmicensis gemachet." Inventar 1587, fol. 313v.
71. When Philipp Hainhofer saw the Lüneburger Spiegel in 1628, he recognized the crowning figure as the statue that appeared before Nebuchadnezzar and wrote, "Die statua Danielis oder der traum des Nebucadnezars, samt derselben explication, als den vier Monarchien, underschidner Reichwappen, und andern alls Köstlich von getribner arbeit, so vil tausent gulden costet mit gar schönen stainen aus fleissigst Geziert mit grosser verwunderung zu sehen und zu Lünburg solle sein gemacht worden." Hainhofer, "Des augsburger Patriciers Philipp Hainhofer Reisen nach Innsbruck und Dresden," 10.
72. The personification of Vice is rendered in the same position in an engraving by Johannes Sadeler after Martin de Vos from 1579, titled *Politeia: Constituit rerm formas Politeia serna. Si bona da Regem, que si perversa Tirannum*. See Martin Warnke, "Die Demokratie zwischen Vorbildern und Zerrbildern," in *Zeichen der Freiheit*, ed. Gamboni, Germann, and Capitani, 84.
73. Schröder, *Dirich Utermarke ein hamburger Goldschmied der Renaissance*, 96.
74. The same figures in the exact arrangement appear in Isaac Schwendter's *Das gute Regiment der Reichsstadt Regensburg* from 1592, which still hangs in the Altes Rathaus in Regensburg, and again in the frontispiece to Georg Lauterbeck's *Regentenbuch*, published in Frankfurt in 1600. See Adolf Schmetzer, "Die Restaurierung des Reichssaales," in Regensburg, *Das Rathaus zu Regensburg*, 174.
75. For a discussion of the role of Justice in allegories of good government, see Polzer, "Ambrogio Lorenzetti's 'War and Peace' Murals Revisited," 77–81.
76. For an excellent discussion of the ways mirrors solicit the gaze of viewers, see Susan L. Smith, "The Gothic Mirror and the Female Gaze," in *Saints, Sinners, and Sisters*, ed. Caroll and Stewart, 74.

Chapter 4

1. Lamalle and Trigault, *La propagande du P. Nicolas Trigault*; Chapuis and Droz, *Automata*, 77–84; Klaus Maurice, "Propagatio fidei per scientias: Jesuit Gifts of Clocks to the Chinese Court," in *The Clockwork Universe*, ed. Maurice and Mayr, 33–34; Riskin, "Machines in the Garden."
2. The annual donation continued until 1687, with the exception of the years between 1632 and 1652, when ducal funds were tied up in the Thirty Years' War. Georg Leidinger, "Herzog Wilhelm V. von Bayern und die Jesuitenmission in China," *Forschung zur Geschichte Bayerns* 12 (1904): 171–75.
3. "Tertium est in centauri forman expressum qui supra mensam seipso discurrens, sagittam emittit ita fortiter ut etiam ligno vicino infigi possit; horas autem ungula pulsat et ferociter caput circumvolvit ad singula momenta aliaque nonulla exhibet quae mihi exciderunt." Lamalle and Trigault, *La propagande du P. Nicolas Trigault*, 104.
4. Spence, *The Memory Palace of Matteo Ricci*, 180–84. On the Jesuits' clockwork gifts in China, see also Ricci, *China in the Sixteenth Century*, bk. 1, chap. 4, and bk. 4, chap. 12. Pfister, *Notices biographiques et bibliographiques sur les Jésuites de l'ancienne mission de Chine*, notice 88; Bailey, *Art on the Jesuit Missions in Asia and Latin America*, 124; and Chapuis and Droz, *Automata*, 315.
5. According to the 1598 inventory of the ducal Kunstkammer in Munich and subsequent descriptions of the collection, this peculiar object was situated on a table in the northeast corner of the chamber, just outside the *Gewölbe* (vault) that housed the most valuable objects in the collection. On the contents of the Gewölbe, see Seelig, "Die münchner Kunstkammer," 104–5. Johannes Fickler recorded the object in the 1598 inventory of the Kunstkammer: "Auf einem uberlengeten stöckl, von Hebeno ein grien geschmelzte berg, darauf ein viereckhet gulden gestell in die vierung mit rundafelin versezt, auf welchem sizt ein Aff von gold, geschmelzt, mit einer blawen Kappen, und ein Paketen in der handt, vor im ein gulden Pulpit, daruaf ein gesangbuech, der wirt von einem urwerckh bewegt, das der Aff mit der Paketen die Mensure schlegt. Neben dem Affen ligen ain Hirsch, ain stuckh wildt von goldt, waiss geschmelzt, ain gulden Rech. Oberst neben dem berg ain anderer geschmeltzer baum. Auf der seitens des stöckhels würt ein Deckhel fürgeschoben, darunder ein täfel in dem ein wald, darinen hirschen und Rech, von minature gemalt, in mitten am fürschub ain Porten an deren baiderseits ain Pyramis mit rubin am fuess versezt, oben auf geschmelzt, am Spiz ein Perlin hinder iedem ein Pyramis ain gulden turlen, auf der objerseitten zwai schubladen, so mit gulden geschmelzten hirsköpfl heraussgezogen werden. In dem grossern ligen ain guldene Reichpfeiffen, auch in dem clainern schublädl, das ein falsch, an idern ein geschmelzt hirshköpfl. Das fuetral hiezu ist inwendig mit weissem Attlass gefüertet, von aussen mit Plawen sammet uberzogen." Diemer, Bujot, and Diemer, eds., *Johannes Baptist Fickler das Inventar der Münchner herzoglichen Kunstkammer von 1598*, no. 3390. See also Diemer, Diemer, and Seelig, eds., *Die münchner Kunstkammer*, vol. 2, part 2, 1057–58; Hoyer and Stierhof, eds., *Schatzkammer der Residenz München*, Inv. no. 609-162; Lietzmann, *Valentin Drausch und Herzog Wilhelm V. von Bayern*, 76–78.
6. Duke Albrecht V began collecting in earnest around 1550. By 1557 his councilors were voicing their disapproval of his use of the duchy's funds. The memorandum that Albrecht's advisors issued is reproduced in Sigmund Riezler, "Zur Würdigung Herzog Albrechts V. von Bayern," *Abhandlungen der bayerischen Akademie der Wissenschaften*, 1894, 67–132. See also Pilaski, *The Munich Kunstkammer*, 2.
7. On the procurement of gems and jewelry for the Kunstkammer and the court, see Stockbauer, *Die Kunstbestrebungen am bayerischen Hofe unter Herzog Albrecht V. und seinem Nachfolger Wilhelm V.*, 91–107.
8. Diemer, Bujot, and Diemer, eds., *Johannes Baptist Fickler das Inventar der Münchner herzoglichen Kunstkammer von 1598*, no. 156.
9. Hsia, *Social Discipline in the Reformation*, 40.
10. Dieter Albrecht, "Bayern und die Gegenreformation," in *Beiträge zur bayerischen Geschichte und Kunst*, ed. Glaser, 13.
11. Claus-Jürgen Roepke, "Die evangelische Bewegung in Bayern im 16. Jahrhundert," in *Beiträge zur bayerischen Geschichte und Kunst*, ed. Glaser, 103.
12. Brecht, *Martin Luther*, 2:88–89.
13. Peter Matheson, introduction to *Argula von Grumbach*, 21–24.
14. On book censorship in Bavaria throughout the sixteenth century, see Heigel, "Die Censur in Altbaiern," 5–32.
15. As quoted in Hsia, *Social Discipline in the Reformation*,

91. In 1558 Albrecht V also appointed the uncompromising Simon Eck to serve as his personal chancellor. Throughout Eck's tenure as chancellor he urged Albrecht to ignore the pleas of the Protestants and align himself more closely with Rome and the Jesuits. See Goetz, *Die bayerische Politik im ersten Jahrzehnt der Regierung Herzog Albrechts V. von Baiern,* 110–15.

16. Roepke, "Die evangelische Bewegung in Bayern im 16. Jahrhundert," 105.
17. Rössler, *Geschichte und Strukturen der evangelischen Bewegung im Bistum Freising 1520–1571*, 176.
18. Hsia, *Social Discipline in the Reformation*, 45.
19. J. C. Smith, *Sensuous Worship*, 3.
20. Katharina Pilaski, "Der Jesuit im Nierenstein: Eine Merkwürdigkeit aus dem Wittelsbacher Hausarchiv," *Aviso: Zeitschrift für Wissenschaft und Kunst in Bayern* 4 (2005): 20–21.
21. Thomas, *A House Divided*, 2.
22. On the Wittelsbachs' patronage during the Catholic Reformation, see J. C. Smith, *Sensuous Worship*, 57–75.
23. As quoted in Thomas, *A House Divided*, 51. On the objects in the Kunstkammer, and for the inventory entry recording Cavalieri's text, see Diemer, Bujot, and Diemer, eds., *Johannes Baptist Fickler das Inventar der Münchner herzloglichen Kunstkammer von 1598*, no. 56. On the portraits, see Reber, *Die Bildnisse der herzoglich bayerischen Kunstkammer nach dem Fickler'schen Inventar von 1598*, 2–56.
24. Diemer, Bujot, and Diemer, eds., *Johannes Baptist Fickler das Inventar der Münchner herzoglichen Kunstkammer von 1598*, nos. 510, 556, 558(1), 558(3), 558(4).
25. Ibid., nos. 2097, 2098, 2101.
26. Ibid., no. 75.
27. There is only one instance in the literature from the period that places a monkey in the role of the hunter. It occurs in Georg Rollenhagen's *Froschmeuseler*, book II, part 2, chap. 11, vv. 125–54, in the passage "Der jagende Affe muss sich den Spott der Krähe gefallen lassen, weil seinem Bogen die Sehne fehlt." As quoted in Dicke, *Die Fabeln des Mittelalters und der frühen Neuzeit*, 25.
28. On émail ronde bosse technique, see Müller and Steingräber, "Die französische Goldemailplastik um 1400"; and Renata Eikelmann, "Goldemail um 1400," in *Das goldene Rössl*, ed. Baumstark, 106–9.
29. To be sure, the iconographic identification of the ape on this object is not clear-cut. The pointer, which the animal holds in its right hand, could suggest that the animal is referring to a pendant scholar. Satiric texts on pendant scholars were widespread by the end of the sixteenth century. If the figure can be construed as a pedant, then the claim of the chapter that this object is essentially about the breakdown of communication still holds. On views of the pendant scholar in early modern Germany, see Kivistö, *The Vices of Learning*, 8–27.
30. On the iconography of the stag and its association with the sense of hearing, see Hassig, *Medieval Bestiaries*, 40–52, and Bath, *The Image of the Stag.*
31. Vitruvius, *De Architectura*, 2:105.
32. "Ain schwartzer berg auss lapide elidio, darob sitz ein af mit eine Musich buch vor Ihme, der schlegt den tact, und rühret die augen, umb Ihn hero sitzten etliche their alle gulden und geschmelzt, sihet als wann der Wolf den gänsen predigte." Häutle, "Die Reisen der Augsburgers Philipp Hainhofer nach Eichstädt, München und Regensburg in den Jahren 1611, 1612 und 1613," 95.
33. By the time the automaton was manufactured, the tale of the guileful fox had been one of the most widespread stories north of the Alps for three hundred years. The first known text to describe Reynard's exploits, *Ysengrimus*, a Latin epic poem over 6,500 lines in length, is believed to have been written in Ghent in 1148–49. The first vernacular version of Reynard's travails was penned in France and titled *Roman de Reynard.* Between 1150 and 1190 the stories of Reynard flourished. The renditions of Reynard's exploits spread in all directions, and we have evidence of manuscripts dating to this forty-year period produced in territories now known as Belgium, Holland, northern Germany, France, and even northern Italy. In 1191 the Alsatian Heinrich der Glichesaer produced the first German version, *Reinhart Fuchs.* Shortly thereafter, an Italian version of the epic appeared, *Rainaldo e Lesegrino.* In the 1260s a Frenchman known as Rutebeuf wrote *Renart le Bestournée.* Sometime between 1263 and 1270 a third French version of the epic appeared. This anonymous text was titled *Couronnement de Renart.* Sometime between 1288 and 1289 Jacquemart Gilée drafted *Renart Reynard le Nouvel.* The first Dutch version, *Van de vos Reynaerde*, was drafted at the end of the thirteenth century. Three more Dutch versions followed in the fourteenth century: *Renyanaerts Historie*, *Renaert de Vos*, and *Hystorie van Renaert die Vos.*

The tale was also well known in England. Chaucer devoted 680 lines to the fox in *The Nun's Priest's Tale of the Cook and the Fox*, and one hundred years later William Caxton published the *Historye of Renaert the Foxe*. A number of studies address the history and development of the Reynard tale. See A. J. Barnouw, "Reynard the Fox," in *Reynard the Fox and Other Mediaeval Netherlands Secular Literature*, ed. Colledge, 47–157; Blake, "English Versions of Reynard the Fox in the Fifteenth and Sixteenth Centuries"; Dicke, *Die Fabeln des Mittelalters und der frühen Neuzeit*, 698–99; Honegger, "A Fox Is a Fox Is a Fox."

34. One scholar has argued that the images of apes satirizing members of the cloth "are likely to have been intended as satires on unworthy individuals who performed these rituals just as ape-knight and ape-physicians almost certainly constituted criticism of men who were not good members of their profession rather than a criticism of the professions themselves." Karl P. Wentersdorf, "The Symbolic Significance of the Figurae Scatologicae in Gothic Manuscripts," in *Word, Picture, and Spectacle*, ed. Davidson, 3.
35. Aristotle, *Historia Animalium*, 2:8, and Aristotle, *Poetics*, 1461b:20.
36. Isidorus of Seville, *The Etymologies of Isidore of Seville*, 253.
37. As quoted in Janson, *Apes and Ape Lore in the Middle Ages and the Renaissance*, 337.
38. Rabelais, *Gargantua and Pantagruel*, 103–4.
39. Ibid., 103. On the relationship between reason and "Becoming Animal" in the early modern period, see Fudge, *Brutal Reasoning*, 59–83.
40. Marginalia were also referred to as *babuini* (monkey business) during the late Middle Ages. Camille, *Image on the Edge*, 12. For an excellent catalog and overview of apes in the margins of medieval manuscripts, see Randall, *Images in the Margins of Gothic Manuscripts*.
41. Camille, *Image on the Edge*, 12.
42. Other languages have similar terms. For instance, imagery and literature that portray an inverted world can fall under the headings of "Mundus Perversus," "Mondo alla Rovescio," "Monde à l'envers," "Mundo al Revès," and "Verkeerde Wereld." See David Kunzle, "The World Upside Down: The Iconography of a European Broadsheet Type," in *The Reversible World*, ed. Babcock, 41.
43. Many of the zoological treatises were derived from medieval bestiaries such as the early Christian *Physiologus*, Isidorus of Seville's *Etymologiae*, Ambrose's *Hexaemeron*, the anonymous *De Bestiis et aliis rebus*, Hugh of Fouilly's *De avibus*, and Solinus's *Collectanea rerum memorabilium*. See Hassig, *Medieval Bestiaries*, 5.
44. Brant, *Das Narrenschiff* (Basel, 1494), and Sachs, *Sämtliche Fabeln und Schwänke*, vols. 1–3. The passage where Sachs discusses a peasant drinking a spoon reads as follows: "Ein dorff in einem Bauren fasas, Der gerne leffel mit milch as Sampt einen grossen wecke, Viers hauser hat sein ecke, vier wagen spandt erfur sein pferdt, Sein küch stundt mitten in derm her, Bold stadel was wein hewe, sein hoff lag in dem strewe, sein stall findt mitten in derm Ross, Sein offen in dabl rod er Schoss, Auss kes macht er gutt milsche, Bon jppen war sein milche, er seblug die bau aus der gruben, und seldtacker auss den Rubern, mit garbe Trölcht er slegel, Auff der spitz stellt ein kegel." Sachs also describes a peasant's paradise where no one works and people are paid to sleep. The streets are lined with plum pudding, houses are made from cake, and food falls randomly from the sky—a culinary utopia. One among many key examples of the world turned upside down is an instance in the text when animals invite humans to devour them. "Auch fliegen umb, muget iyr glauben / Gebraten Hüner, Gensz, und Tauben / Wer sie nigh fact und ist so faul / Dem fliegen sie von selbs in da maul / Die sew all Jar gar wol gerathen / Lauffen im Land umb, sind gebraten! / Yede eyn messer hat im Rück!"
45. The term *misericord* is derived from the Latin *misericordia*, which denoted mercy or pity. Misericords are carved in wooden brackets beneath the seats of choir stalls in cathedrals or collegiate and monastic churches. Grössinger, *The World Upside-Down*, 11–17.
46. Hayward, *Virtuoso Goldsmiths and the Triumph of Mannerism*, plate 555. Art historians have been curiously silent with regard to the luxury items that depict the *Verkehrte Welt*. For this literature, see Natalie Zemon Davis, "Women on Top: Symbolic Sexual Inversion and Political Disorder in Early Modern Europe," in *The Reversible World*, ed. Babcock, 154; Jonassen, "Lucian's 'Saturnalia,'" 58–68; Moxey, *Peasants, Warriors, and Wives*, 45–62; and Kunzle, "The World Upside Down," 40.
47. See Kunzle, "The World Upside Down," 39–94.

48. The exception that proves the rule is Swan, *Art, Science, and Witchcraft in Early Modern Holland*, 157–74.
49. Quintilian, *Institutiones*, 8.6.54.
50. Grassi and Lorch, *Folly and Insanity in Renaissance Literature*, 67.
51. Keith Moxey, "Hieronymus Bosch and the 'World Upside Down': The Case of *The Garden of Earthly Delights*," in *Visual Culture*, ed. Bryson, Holly, and Moxey, 104–40.
52. *The Oxford English Dictionary*, 2d ed. (Oxford: Oxford University Press, 1989).
53. Robert W. Scribner, "Demons, Defecation and Monsters: Popular Propaganda for the German Reformation," in Scribner, *Popular Culture and Popular Movements in Reformation Germany*, 278–99.
54. On the propagandistic role of these images, see Scribner, *For the Sake of Simple Folk*, esp. chap. 7, "Teaching the Gospel: Propaganda as Instruction," 190–228.
55. On the pedagogical function of the printed catechisms, see Strauss, *Luther's House of Learning*, 151–75.
56. The large two-block print survives in two versions. One is supplemented with text composed by Melancthon's most fervent interlocutor, the theologian Matthias Flacius Illyricus (1520–1575); the other is without commentary. The authorship of this print has been questioned. Because of its manner and theme the work has been attributed to Cranach's workshop, but the watermark suggests that it was printed in Magdeburg, where Illyricus was preaching and writing. Since we know that Kempf produced satirical prints in Magdeburg that supported the reformers' cause, it has been claimed, and for the most part accepted, that Kempf executed the woodcut in Magdeburg under the guidance of Illyricus. I thank Suzanne Karr Schmidt for relaying this information to me.
57. This representation of the demon imparting the teachings of Satan to the monk's ear is an inversion of the popular legend of Saint Gregory the Great. After Gregory's death it was suggested by his contemporaries that his works were heretical and should be burned. One of Gregory's followers prevented the book burning by testifying that he had witnessed the Holy Spirit, in the form of a dove, dictating the works in Gregory's ear. See Westfehling, *Die Messe Gregors des Grossen*, 13.
58. That this woodcut underscores the moral, spiritual, and theological oppositions of the Reformed preacher and the Catholic monk aligns it with the "Wittenberg Tradition," developed from Luther's writings, of the antichrist. Popular depictions of the antichrist, such as the monk as the devil himself or as being born from a she-devil, are surprisingly not present in this work. On representing the pope as the antichrist, see Scribner, *For the Sake of Simple Folk*, 155–63.
59. As quoted in Certeau, *The Mystic Fable*, 112.
60. Koerner, *The Reformation of the Image*, 137–52, 252–81.
61. Luther, "The Office of Preaching and of Preachers and Hearers," in *The Sermons of Martin Luther*, 3:374.
62. Ibid., 3:376.
63. As quoted in Strauss, *Luther's House of Learning*, 4.
64. Ibid., 51:11.
65. Harold Dwight Lasswell, "The Structure and Function of Communication in Society," in *The Communication of Ideas*, ed. Bryson, 148.
66. Robert Scribner, "Oral Culture and the Diffusion of Reformation Ideas," 238.
67. In addition to charismatic figures such as Hus and von Kaysersberg, numerous itinerant preachers predicted the end of days to rapt audiences. The phenomenon was so widespread that in 1513 the Lateran Council prohibited preaching on such matters. Cunningham, *The Four Horsemen of the Apocalypse*, 1.
68. Louth, "Augustine on Language." See also Tanner and Watson, "Least of the Laity"; Hamm, "Normative Centering in the Fifteenth and Sixteenth Centuries," 315–23; and Haas, *Sermo mysticus*.
69. These crucial aspects of devotion in the late medieval church have been called *heilbringende Schau* (salvific display) by A. L. Mayer. See Mayer, "Die heilbringende Schau in Sitte und Kult," in Casel, *Heilige Überlieferung*. For a more recent and seminal work on the topic, see Hamm, "Frömmigkeit als Gegenstand theologiegeschichtlicher Forschung."
70. One commentary on the Canon reads: "It is unseemly that the laity should be concerned with these things." Quoted by Koerner in *The Reformation of the Image*, 352.
71. As quoted in ibid.
72. O'Malley, *Praise and Blame in Renaissance Rome*, 36.
73. On Jesuits' practices of preaching, see O'Malley, *The First Jesuits*, 91–103.
74. For an excellent discussion of this and other prints of clerical figures in the guise of the red dragon and other beasts of the apocalypse, see Christiane Andersson, "Polemical Prints in Reformation

Nuremberg," in *New Perspectives on the Art of Renaissance Nuremberg*, ed. J. C. Smith, 48–59.

75. I am borrowing this phrase from Joseph Leo Koerner. See Koerner, *The Moment of Self-Portraiture in German Renaissance Art*, 60.
76. Scribner, *For the Sake of Simple Folk*, 11.

Chapter 5

1. In official Habsburg papers the term *munus honorarium* was employed, as well as *Ehrengeschenk* (honor-gift or honorarium). See Johnson, *Cultural Hierarchy in Sixteenth-Century Europe*, 222. *Verehrung* is the modern German spelling of the term; the late middle-German spelling is *vereegung*. In certain contexts, the term *verehrung* is also a synonym for reverence (*reverentz*).
2. A wide range of linguistic, ethnic, and religious groups made up the population of the early modern Ottoman Empire. Members of the ruling elite of the Ottoman Empire referred to themselves as "Ottomans" (*Osmanli* in Turkish), after the founder of the dynasty, Osman. "Turk," in classical Ottoman, was used pejoratively to denote inhabitants of the empire. German-speaking peoples in the Holy Roman Empire referred to all inhabitants of the Ottoman Empire as "Turks" or *Türken*. In this chapter I have followed the lead of Margaret Meserve and use the terms "Ottoman" and "Turk" interchangeably to underscore an assumption of German-speakers in the Holy Roman Empire in the sixteenth century: that all "Turks" were members of an identifiable race, whose lineage could be traced back to antiquity. See Meserve, *Empires of Islam in Renaissance Historical Thought*, vii.
3. Kurz, *European Clocks and Watches in the Near East*; Gottfried Mraz, "The Role of Clocks in the Imperial Honoraria for the Turks," in *The Clockwork Universe*, ed. Maurice and Mayr, 37–48; J. Michael Rogers, "The Gorgeous East: Trade and Tribute in the Islamic Empires," in *Circa 1492*, ed. Levenson, 69–78; Jardine, *Worldly Goods*; Raby, "The Serenissima and the Sublime Porte: Art in the Art of Diplomacy," in *Venice and the Islamic World, 828–1797*, ed. Carboni, 90–119.
4. Many automata listed in diplomatic accounts of imperial ambassadors who traveled to Istanbul display scenes of Turks in ceremonial processions as well. See Loebl, *Zur Geschichte des Türkenkrieges von 1593–1606*, appendixes A and B. The objects are now referred to as the *Sultan on Horseback* (ca. 1570, Dresden, Mathematisch-Physikalischer Salon); the *Sultan on Horseback with Attendants* (ca. 1580, Vienna, Kunsthistorisches Museum); and the *Pashas on Horseback* (ca. 1585, Newark, Newark Museum of Art). That gifts intended for the Ottomans remained in collections in the Holy Roman Empire is attested by the 1578 inventory of Holy Roman Emperor Maximilian II's possessions, which records two watches "von der Türckhischen Verehrung uberblieben" along with several silver vessels and a clock with a glass case. *Jahrbuch der Kunsthistorischen Sammlungen des Allerhöchsten Kaiserhauses* 13 (1892): xci, nos. 176, 490, 623.
5. Maurice, *Die deutsche Räderuhr*, 667–70; and Gottfried Mraz, "The Role of Clocks in the Imperial Honoraria for the Turks," 37–48.
6. Kafadar, *Between Two Worlds*, 20.
7. For recent analyses of European representations of Turks and how those representations spoke to the interests, assumptions, and fears of their early modern European makers and audiences, see Harper, ed., *The Turk and Islam in the Western Eye*.
8. The literature on European gifts to the Ottoman court is vast. For an extensive bibliography, see Carboni, ed., *Venice and the Islamic World*, 350–69.
9. Mraz, "The Role of Clocks in the Imperial Honoraria for the Turks," 40.
10. Imperial account books note that during the first year of the Türkenverehrung the empire sent 72,709 gulden and 13 kreuzer to the sultan and 1,000 ducats and 3 vessels, made entirely of silver, to the pasha in Ofen (Budapest). This amount of money given remained relatively stable throughout the years of the Türkenverehrung. For instance, in 1549 the sultan received 72,779 gulden and 35 kreuzer; in 1550, 68,508 gulden and 16 kreuzer. Mraz, "The Role of Clocks in the Imperial Honoraria for the Turks," 39. On average, four silver gulden were the equivalent of one golden ducat, and seventy-two kreuzer equaled one gulden. See "Gulden," in *Meyers kleines Konversations-Lexikon*, 7:922.
11. Maurice and Mayr, eds., *The Clockwork Universe*, 23.
12. "Ain lange rolln wie der Türggisch Kaiser geen Kurchen reüt" and "Ein buch von der hand gemalt, sein allerley türkische trachten und ire *cermonien*." *Jahrbuch der Kunsthistorichen Sammlungen des aller höchsten Kaiserhauses* 7 (1888): ccxl; and Bauer and Haupt,

eds., "Das Kunstkammerinventar Kaiser Rudolfs II.," fol. 382, no. 2713.

13. Loebl, *Zur Geschichte der Türkenkrieges von 1593–1606*, 117.
14. Terzioğlu, "The Imperial Circumcision Festival of 1582," 82–84.
15. On Donatello's equestrian statue, see Pope-Hennessy, *Italian Renaissance Sculpture*, 53–54. On the figures at Varallo, see Alessandro Nova, "Popular Art in Renaissance Italy: Early Response to the Holy Mountain at Varallo," in *Reframing the Renaissance*, ed. Farago, 113–26. For Maximilian's tomb, see Wood, *Forgery, Replica, Fiction*, 314–15.
16. The processions of the Ottoman sultans changed very little over the sixteenth century. The ceremonies were organized according to the procession registers kept in the palace archives, and the master of ceremonies made sure that all the protocol guidelines were strictly followed. See Nuhran Atasoy, "Processions and Protocol in Ottoman Istanbul," in *The Sultan's Procession*, ed. Ådahl, 169–70.
17. Necipoğlu, *Architecture, Ceremonial, and Power*, 25.
18. Ibid., 30.
19. Terzioğlu, "The Imperial Circumcision Festival of 1582," 86.
20. Curipeschitz, *Itinerarium der Botschaftsreise des Josef von Lamberg und Niclas Jurischitz durch Bosnien, Serbien, Bulgarien nach Konstantinopel*; Ramberti, *Libri tre delle cose de Turchi*; Beg and Gritti, *Opera nova la quale dechiara tutto il governo del gran turcho*; Geuffroy, *Briesve description de la court du Grant Turc, et ung sommaire du règne des Ottmans*; Busbecq, *The Turkish Letters of Ogier Ghiselin de Busbecq*; Schweigger, *Ein newe reyssbeschreibung auss Teutschland nach Constantinopel vnd Jerusalem*; and Gerlach, Gerlach, Wagner, and Friese, *Stephan Gerlachs dess aeltern Tage-Buch*.
21. On printed processions, see Larry Silver, "Triumphs and Travesties: Printed Processions of the Sixteenth Century," in *Grand Scale*, ed. Silver and Wyckoff, 15–32.
22. For an overview of the composite print phenomenon in the sixteenth century, see Appuhn and Heusinger, *Riesenholzschnitte und Papiertapeten der Renaissance*. For a more recent account, see Silver and Wyckoff, eds., *Grand Scale*.
23. It should be noted that Coecke van Aelst was not the first artist to produce a composite print of a parade of Turks. Between 1529 and 1530 Erhard Schön produced a frieze of fifteen prints, accompanied by the text of Hans Sachs, depicting the atrocities of the Turks after the siege in Vienna. See Moxey, *Peasants, Warriors, and Wives*, 76–77.
24. Karel van Mander claims that Coecke van Aelst was in Istanbul to negotiate for a sale of tapestries to the sultan on behalf of Willem Dermoyen of Brussels. See Astrit Schmidt-Burkhardt, "Pieter Cocke van Aelst: Sitten und Gebräuche der Türken," in *Europa und der Orient, 800–1900*, ed. Sievernich and Budde, 137; Necipoğlu, "Süleyman the Magnificent and the Representation of Power in the Context of Ottoman-Habsburg-Papal Rivalry," 419–20. See Jardine and Brotton, *Global Interests*, 82–87; and T. P. Campbell, *Tapestry in the Renaissance*, 251–52.
25. On the Ottomans' relationship to the past of Istanbul, particularly its Greek past, see Meserve, *Empires of Islam in Renaissance Historical Thought*, 67.
26. On problems of representing a *punctum temporis* in painting, see Gombrich, "Moment and Movement in Art."
27. Alexandrine St. Clair identified the Fatih Mosque in the left background, the Mosque of Firuz Aga in the middle ground, and the serpentine column and Obelisks of Theodosius and Constantine VII in the foreground. St. Clair, *The Image of the Turk in Europe*, cat. no. 5.
28. Please note that here I am focusing on images of Ottoman rulers. There are certainly other attitudes, hostile and benign, expressed in images of Turks more generally in the Holy Roman Empire and Europe. For instance, see Raby, *Venice, Dürer, and the Oriental Mode*; Aikema and Brown, eds., *Renaissance Venice and the North*; Mellinkoff, *Outcasts*. For a more recent treatment of Turks in Northern European print culture, see Larry Silver, "East Is East: Images of the Turkish Nemesis in the Habsburg World," in *The Turk and Islam in the Western Eye*, ed. Harper, 185–215.
29. Bronwen Wilson makes this persuasive argument in two publications: "Reflecting on the Turk in Late Sixteenth-Century Venetian Portrait Books"; and *The World in Venice*, esp. chap. 4, "Reproducing the Individual: Likeness and History in Printed Portrait Books," 222–55.
30. Esin Atil, "The Image of Süleyman," in *Süleyman the Second*, ed. İnalcık and Kafadar, 334.
31. "In the period that we are dealing with, Ottoman peace agreements with other nations were considered by the Ottomans to be truces rather than

bilateral treaties. Since a continuous status of peace with infidels is not permissible according to classical Islamic principles, it was the usual practice for the Ottomans to conclude a temporary truce of ten, twenty, and even thirty years." Bülent Ari, "Early Ottoman Diplomacy: Ad Hoc Period," in *Ottoman Diplomacy*, ed. Yurdusev, 37. Mattingly, *Renaissance Diplomacy,* 101–7.

32. Habsburgisch Ungarn stretched from Dubrovnik in the south to the tributary of the Vistula (in German, the Weichsel) river in the north, and across Bratislava in the west to present-day Miskolc in the east. The Königreich Böhem encompassed modern Austria, Slovakia, and the Czech Republic. These regions correspond to almost the entire region of the Balkans, with the exclusion of present-day Croatia, and the western half of present-day Ukraine. Consequently, the central plains of Hungary were under the control of the Ottomans, who organized them into frontier provinces (*beglergegilik*). İnalcık and Quataert, eds., *An Economic and Social History of the Ottoman Empire*, 304.
33. The ducat was the standard gold coin throughout the Empire. It was not, however, imperially sanctioned until 1566. "Gulden," in *Meyers kleines Konversations-Lexikon*, 7:922.
34. Goffman, *The Ottoman Empire and Early Modern Europe*, 73.
35. Loebl, *Zur Geschichte der Türkenkriege von 1593–1606*, 44.
36. See Jardine, *Worldly Goods*, esp. chap. 2, "The Price of Magnificence," 91–132.
37. Cutler, "Gifts and Gift Exchange as Aspects of the Byzantine, Arab, and Related Economies," 251. On the regularity of the practice of gift giving in an early modern diplomatic context, see Mattingly, *Renaissance Diplomacy*, 101–7.
38. Anderson, *The Rise of Modern Diplomacy*, 16.
39. The latter portion of the ambassador's duties was taken seriously. The imperial representative was required to write reports to the Holy Roman Emperor no less frequently than every four days. See Gévay, ed., *Urkunden und Actenstücke zur Geschichte der Verhältnisse zwischen Österreich, Ungern und der Pforte im XVI. und XVII. Jahrhunderte aus Archiven und Bibliotheken*, 1:i–iv.
40. Mattingly, *Renaissance Diplomacy*, 98.
41. Anderson, *The Rise of Modern Diplomacy*, 10. Süleyman did send an ambassador to the Holy Roman Empire to witness Maximilian II's coronation as King of the Romans in November 1562. At that time the envoy, Ibrahim Bey, presented to Holy Roman Emperor Ferdinand I a draft of the renewed peace treaty and gifts, which included crystal vessels, camels, and horses. Johnson, *Cultural Hierarchy in Sixteenth-Century Europe*, 197.
42. It was not until the Habsburgs and the Ottomans signed a peace treaty at Zsitva-Török in 1606 that the Habsburg ruler was considered equal to the Ottoman Sultan. İnalcık and Quataert, eds., *An Economic and Social History of the Ottoman Empire*, 423.
43. Busbecq, *The Turkish Letters of Ogier Ghiselin de Busbecq*, 39.
44. Callières, *The Art of Diplomacy*, 56.
45. *Jahrbuch der Kunsthistorischen Sammlungen des Allerhöchsten Kaiserhauses* 7 (1888): cxix, reg. 4980.
46. Ibid., cxix, reg. 4991.
47. Ibid., cxxvii, reg. 5113.
48. Kugler and Seipel, *Kaiser Ferdinand I*, 302.
49. "Dem Nassadistch Obersten Andreas Tarnoczj zu Komorn, welche für dem Pascha von Ofen als Geschenke bestimmt sind als: ein silbernes und vergoldetes doppeltes Trinkgeschirr von 4 mark 12 Loth Wiener Gewicht, das nebst dem dazuhörigen Futteral von mert Pappierer, Goldschmied zu Wien, geliefert wurde, weiters ein Panzerhemd, wleches von Karl Schwetkhowicz um 175 Gulden rheinisch gekauft wurde. Für Sammtliche Objeckte erscheinen 270 Gulden 35 Kreuzer rheinisch in Ausgabe"; and "Hanns Runckl, Uhrmacher zu Augsburg, erhält für eine Uhr, welcher Kaiser Maximilian II. nachträglich einer Person in Constsantinople verehrte, 38 Gulden." *Jahrbuch des Kunsthistorischen Sammlungen* 7 (1888): cv, reg. 4852 and cxxi, reg. 5019.
50. Busbecq, *The Turkish Letters of Ogier Ghiselin de Busbecq*, 131.
51. Ibid., 62.
52. On the Venetian gifts, see Raby, "The Serenissima and the Sublime Porte," 94–96.
53. Cutler, "Significant Gifts," 91–93.
54. "Ob wol diese Credenz und Silbergeschirr alles zumal von sehr köstlicher Arbeit / die wol höher möcht geacht werden / dann das Metall / das Gold oder Silber / so gilt es doch nichts bey diesen Leuten / sie verwundern sich zwar drüber / aber wie man mich bericht / so lassen sie es alles wider schmelzen / und machen Münss oder Gelt daraus / die schönen Uhrenwerck soll der Sultan in einem grossen Gemach auff einem hauffen stehn haben /

die ihm von vielen Jahren her seyn zukommen / die verderben von dem Rost / etlich werden verkaufft / bissweilen lest er abwechsseln / und ihm eines derselben ins Gemach stellen / wenn ers nun ein gute Zeit gebraucht/ lest ers hinweg thun / und ein anders herfür bringen / und also fort." I have used Claudia Swan's translation of this passage. She discusses Schweigger's disappointment at the treatment of the tribute in an unpublished lecture, "'*Rarieteyten van dese Landen*': Transcultural Lives of Early Modern Exotica." An alternative translation can be found in Hedda Reindl-Kiel, "Ottoman-European Cultural Exchange: East Is East and West Is West, and Sometimes the Twain Did Meet; Diplomatic Gift Exchange in the Ottoman Empire," in *Frontiers of Ottoman Studies*, ed. Imber and Kiyotaki, 113.

55. Terzioğlu, "The Imperial Circumcision Festival of 1582," 86.
56. Çiğ, Batur, and Köseoğlu, *The Topkapı Saray Museum*, 21.
57. Ibid., 23.
58. This claim is resonant with Nancy Bisaha's recent statement on the anachronicity of viewing the relations between the Ottoman Empire and Europe in the early modern period through the lens of Edward Said's *Orientalism*. She writes: "Renaissance humanists, however, present some important challenges to Said's model. Where Said focuses on colonialism as a key component in the formation of the West-East discourse, the bulk of humanist rhetoric on the Turks and Islam shows a highly developed sense of Europe as the cultural superior to the East—precisely at a time when Europe was fighting for its survival." Bisaha, *Creating East and West*, 6.
59. There are many entries in the imperial account books that record payments to Augsburgian artisans for the extra gifts that were sent along with the tribute payment. On February 8, 1566, the documents state: "Dier Kaiserliche Uhrmacher Gerhart Eenmoser erhält für zwei von ihm gefertigte und der Kammer gelieferte uhrlein, welcher Kaiser Maximilian II. durch die kurzlich anwesend gewesene türkische Botschaft nach Constantinople übersendent halte, 68 Gulden rheinisch ausbezahlt." *Jahrbuch des Kunsthistorischen Sammlungen* 7 (1888): cxxi, reg. 5136.
60. It should be noted that a letter written by Pietro Aretino dated October 2, 1531, states that Süleyman brought a gold ring set with clockwork, which was manufactured in Venice. Kurz, *European Clocks and Watches in the Near East*, 22.
61. Mraz, "The Role of Clocks in the Imperial Honoraria for the Turks," 48.
62. Mitrovic, *Adventures of Wenceslas Wratislaw of Mitrowitz*, 1f.
63. Regrettably, only lists from the mid-1560s and onward survive. We therefore cannot say with certainty that automata were given as tribute before this time. We do know through letters and memoirs of Venetian merchants that Ferdinand I began sending elaborate clocks with astronomical and moving figures during the 1540s. According to Paolo Giovio, Ferdinand sent Süleyman a celestial clock with moving figures, so large that twelve men had to carry it. Giovio also believed that the clock had been originally crafted for Holy Roman Emperor Maximilian I. Kurz suggested that the object might have been the famed *Theoria planetarum*, which numerous craftsmen and scholars worked on in Nuremberg during Maximilian I's reign. See Kurz, *European Clocks and Watches in the Near East*, 23, nos. 1 and 2.
64. The travel routes from Vienna to Istanbul were roughly the same for the imperial ambassadors throughout the period of the Türkenverehrung. For more detailed and precise descriptions of the routes, see Gerlach, Gerlach, Wagner, and Friese, *Stephan Gerlachs dess aeltern Tage-Buch*; Schweigger, *Ein newe reyssbeschreibung auss Teutschland nach Constantinopel vnd Jerusalem*; and Busbecq, *The Turkish Letters of Ogier Ghiselin de Busbecq*.
65. Ogier Ghiselin de Busbecq briefly mentions his first encounter with the pasha in his first "Turkish Letter": "We were introduced into the presence of the Pasha, who had recovered from his illness. We tried to mollify him with presents, and then complained of the insolence and misdeeds of the Turkish soldiers and demanded back the places which had been taken from us in violation of the truce and which he had promised in his letter to my sovereign to restore on condition that he sent a representative." Busbecq, *The Turkish Letters of Ogier Ghiselin de Busbecq*, 12.
66. There is a gouache painting (1628) of the count of Kuefstein delivering gifts to the pasha at Ofen at Schloss Greillenstein in lower Austria. See Tietze, *Die Denkmale des politischen Bezirkes Melk*, 482.
67. "In fact, a man who intends to go among the Turks must be prepared, as soon as he has crossed the

frontier, to open his purse and never close it till he leaves the country. Meanwhile he must sow money broadcast and pray that it may not prove unfruitful. If there is no other result, it is at any rate the only method of softening the fierce heart of the Turk, who hates all other nations. Money acts like a charm to sooth [*sic*] their otherwise intractable minds." Busbecq, *The Turkish Letters of Ogier Ghiselin de Busbecq*, 25. On piracy and brigandage in the sixteenth century, see Braudel, *The Mediterranean and the Mediterranean World in the Age of Philip II*, 865–90.

68. Mraz, "The Role of Clocks in the Imperial Honoraria for the Turks," 40.
69. Ibid., 40–41.
70. Kurz, *European Clocks and Watches in the Near East*, 25.
71. On the transmission of technological knowledge by itinerant artisans, see Mathias, "Skills and the Diffusion of Innovations from Britain in the Eighteenth Century"; and more recently, Mokyr, *The Gifts of Athena*.
72. Ben-Zaken, *Cross-Cultural Scientific Exchanges in the Eastern Mediterranean*, 15–21.
73. Çiğ, Batur, and Köseoğlu, *Topkapı Saray Museum*, 33. On England's monopoly of clockwork in the eighteenth century, see Britten and Baillie, *Britten's Old Clocks and Watches and Their Makers*, 77–140.
74. As quoted in Johnson, *Cultural Hierarchy in Sixteenth-Century Europe*, 226.
75. Groebner, *Liquid Assets, Dangerous Gifts*, 1.

Chapter 6

1. Pagani, *Eastern Magnificence and European Ingenuity*, 26–98.
2. For discussions of this commensurability, see Frank, *ReOrient*; Jardine, *Worldly Goods*; Matar, *Turks, Moors, and Englishmen in the Age of Discovery*; Jardine and Brotton, *Global Interests*; Barbour, *Before Orientalism*; and Subrahmanyam, *Courtly Encounters*.
3. Seelig and Spicer, "Die Gruppe der Diana auf dem Hirsch in der Walters Art Gallery."
4. I thank Claudia Swan for notifying me of a Diana Automaton in the inventory of von Sohm's collection. On Rudolf's wax model, see *Jahrbuch der kunsthistorischen Sammlungen* 23 (1905): xxx, no. 491.
5. Seelig and Spicer, "Die Gruppe der Diana auf dem Hirsch in der Walters Art Gallery," 108.
6. Barkan, "Diana and Actaeon," 345.
7. Much has been published on the miniature. See Jeremiah P. Lotsy, "Abu'l Hasan," in *Master Artists of the Imperial Mughal Court*, ed. Pal, 81; Okada, *Indian Miniatures of the Mughal Court*, 54–55, 170–75; and more recently, Bailey, "The End of the 'Catholic Era' in Mughal Painting"; and Ramaswamy, "Conceit of the Globe in Mughal Visual Practice," 754–78.
8. Although none of the albums made for Jahangir remain intact, we have a sense, based on the surviving folios and later albums, of which folios have not been dispersed to collections and how they were organized. The first pages were illuminated with floral motifs, arabesques, or both. The first page was followed by identical facing *shamsas* (a sunburst design). The shamsas were followed by an illuminated preface. Portraits of Mughal emperors came after the preface. The folios that followed the portraits were arranged in groups. Elaine Julia Wright, "An Introduction to the Albums of Jahangir and Shah Jahan," in Wright, *Muraqqa'*, 39–42.
9. The literature is too vast to list here. For a comprehensive bibliography, see Roodenburg and Roeck, eds., *Cultural Exchange in Early Modern Europe*, vol. 4, *Forging European Identities, 1400–1700*; and Thomas DaCosta Kaufmann, "Cultural Transfer and Arts in the Americas," in *The Virgin, Saints, and Angels*, ed. Stratton-Pruitt, 18–25. For exceptions, see note 2.
10. Gruzinski, *Images at War*, 1–5.
11. The literature is vast. For an overview, see Malekandathil, *The Germans, the Portuguese, and India*; Mathew, *Indo-Portuguese Trade and the Fuggers of Germany*; and Boyajian, *Portuguese Trade in Asia under the Habsburgs*. Stephanie Leitch is one of the few art historians to have explored the topic. See Leitch, *Mapping Ethnography in Early Modern Germany*.
12. Lach, *Asia in the Making of Europe*.
13. Subrahmanyam, *The Portuguese Empire in Asia*, 9–28.
14. For discussions of trade and early modern cosmopolitan culture in sixteenth-century Goa, see Pedro Moura Carvalho, "Goa's Pioneering Role in Transmitting European Traditions to the Mughal and Safavid Courts," in *Exotica*, ed. Trnek and Silva; Pereira and Pal, eds., *India and Portugal*; and Flores, ed., *Goa and the Great Mughal*.
15. Subrahmanyam, *Three Ways to Be Alien*, 23–25.
16. Županov, *Missionary Tropics*, 172–80.
17. Santos, *Ethiopia Oriental*.
18. Županov, *Missionary Tropics*, 180
19. B. Cohn, "The Past in the Present," 1–6.

20. Županov, *Missionary Tropics*, 12–15.
21. Maclagan, *The Jesuits and the Great Mogul*, 47.
22. Wink, *Akbar*, 86–108.
23. Fazl, *Akbarnama*, 1027.
24. I believe Monserrate's account of Akbar's adoration of the *Polyglot Bible* reveals more about how Europeans expected and imagined their goods to be received by exotic sovereigns than it does about the reception of Christianity at the Mughal court. The account dwells on how Akbar treated the *Polyglot Bible* as an *object*. He venerates it; he kisses it; he places it on his head; he takes it to a private chamber; he opens and shuts it; and he exhibits it in a "worthy location." Monserrate, *The Commentary of Father Monserrate*, 37.
25. Beach, "The Gulshan Album and Its European Sources."
26. Clifford, *The Predicament of Culture*, 17–19.
27. Subrahmanyam, "A Roomful of Mirrors," 43.
28. Koch, *Mughal Art and Imperial Ideology*, 5.
29. Bailey, *Art on the Jesuit Missions in Asia and Latin America*, 125.
30. Guerreiro, *Jahangir and the Jesuits*, 65.
31. Ebba Koch, "The Baluster Column: A European Motif in Mughal Architecture and Its Meaning," in Koch, *Mughal Art and Imperial Ideology*, 52–57.
32. Beach, "The Gulshan Album and Its European Sources," 73.
33. The literature on Dürer's *Rhinoceros* and Burgkmair's *The Peoples of Africa and India* is extensive. For the *Rhinoceros*, see Schoch et al., *Albrecht Dürer*, 420–24. For *The Peoples of Africa and India*, see Hausberger and Biedermann, eds., *Hans Burgkmair*, cat. nos. 23–26; McDonald, "Burgkmair's Woodcut Frieze of the Natives of Africa and India"; and Leitch, "Burgkmair's *Peoples of Africa and India* (1508) and the Origins of Ethnography in Print."
34. Peter Geffcken, "Die Welser und ihr Handel 1246–1496," in *Die Welser*, ed. Häberlein and Burkhardt, 27–167.
35. Kellenbenz, *Die Fugger in Spanien und Portugal bis 1560*, 1:49–62.
36. Malekandathil, *The Germans, the Portuguese, and India*, 107–8.
37. For details on Cron, see Subrahmanyam, "An Augsburger in Asia Portuguesa: Further Light on the Commercial World of Ferdinand Cron," in *Emporia, Commodities, and Entrepreneurs in Asian Maritime Trade*, ed. Ptak and Rothermund, 401–25.
38. Malekandathil, *The Germans, the Portuguese, and India*, 99–105.
39. Nuno Vassallo e Silva, "Precious Stones, Jewels, and Cameos: Jacques de Coutre's Journey to Goa and Agra," in *Goa and the Great Mughal*, ed. Flores, 126–29.
40. As quoted in Bailey, *Art on the Jesuit Missions in Asia and Latin America*, 124.
41. As quoted in ibid., 123.
42. As quoted in Spence, *The Memory Palace of Matteo Ricci*, 180.
43. Lamalle and Trigault, *La propagande du P. Nicolas Trigault*, 73–75.
44. Chapuis and Droz, *Automata*, 77–84.
45. Bobinger, "Der augsburger Uhrmacher Hans Schlothaim"; Maurice, *Die deutsche Räderuhr*, 2:229–30; Streng, "Augsburger Meister der Schmiedgasse um 1600."
46. On the metalwork from Augsburg in the Wittelsbachs' ducal Kunstkammer in Munich, see Lorenz Seelig, "Die münchner Kunstkammer," in *Die münchner Kunstkammer*, ed. Diemer, Diemer, and Seelig, 3:57–64.
47. Roe, *The Embassy of Sir Thomas Roe to India*, 127.
48. Sir Thomas Roe, "Advise for Goods for Surratt," in ibid., 459.
49. The portrait of Shah Abbas in the Freer miniature is based on a drawing executed by the Mughal artist Bishandas, who traveled to Isafahan as part of a Mughal embassy in 1618. The drawing survives and is housed in the Arthur M. Sackler Museum at Harvard University. On early seventeenth-century portraits of Shah Abbas, see Rizvi, "The Suggestive Portrait of Shah 'Abbas."
50. Jahangir, *The Jahangirnama*, appendix C, 477.
51. Ettinghausen, *Paintings of the Sultans and Emperors of India in American Collections*, 23; Littlefield, "The Object in the Gift," 100–101.
52. Ettingausen, *Paintings of the Sultans and Emperors of India in American Collections*, 23.
53. Rice, "The Emperor's Eye and the Painter's Brush," 159–224.
54. Habsburg, *The St. Petersburg Muraqqa'*, 114.
55. Pinney, "Creole Europe," 131–36.
56. On Heda, see Willigen, *Les artistes de Harlem*, 152–56, 367–71; and Kruijtzer, *Xenophobia in Seventeenth-Century India*, 19–33.
57. Subrahmanyam, "A Roomful of Mirrors," 51.
58. On the way appropriation creates new meanings and values in a new context, see Robert Nelson,

"Appropriation," in *Critical Terms for Art History*, ed. Nelson and Shiff, 164–73; and Ashley and Plesch, "The Cultural Processes of 'Appropriation.'"

59. Most of this scholarship comes from the work of Homi Bhabha, who called attention to the "interstitial passage between fixed identifications" that "opens up the possibility of cultural hybridity that entertains difference without an assumed or imposed hierarchy." Bhabha, *The Location of Culture*, 4.

60. For comprehensive bibliographies on hybridity, see Dean and Leibsohn, "Hybridity and Its Discontents," 32–35; and Flood, *Objects of Translation*, 347–52.

61. Sheldon Pollock, "The Sanskrit Cosmopolis, 300–1300: Transculturation, Vernacularization, and the Question of Ideology," in *Ideology and Status of Sanskrit*, ed. Houben, 246. See also Bhabha, "The Third Space," in *Identity: Community, Culture, Difference*, ed. Jonathan Rutherford, 211; and Subrahmanyam, *Beyond Incommensurability*.

62. Much has been written on this painting. The most comprehensive study of the work is Richard Ettinghausen, "The Emperor's Choice," in *De Artibus Opuscula XL: Essays in Honor of Erwin Panofsky*, ed. Meiss, 98–120. See also Beach, *The Imperial Image*, 168–69; and Beach, "The Mughal Painter Abu'l Hasan and Some English Sources for His Style," 11–14.

63. Sand clocks did exist in the Islamic Middle East and were called *Shishe-ye sa'at* in Persian. The Ottomans also had an industry in Constantinople of hourglass making. However, as Ettinghausen and others have noted, the metal supports on the hourglass in Bichitr's painting are European. Balmer, "The Operation of Sand Clocks and Their Medieval Development"; and on the water clocks at the Mughal court, see Terry, *A Voyage to East-India*, 230–31.

64. Ettinghausen, *Paintings of the Sultans and Emperors of India in American Collections*, 104.

65. Ibid., 108.

66. Rice, "The Emperor's Eye and the Painter's Brush," 115.

67. Ibid., 143.

68. On the copying of English portraits at the Mughal court from 1615 onward, see Ania Loomba, "Of Gifts, Ambassadors, and Copy-Cats: Diplomacy, Exchange, and Difference in Early Modern India," in *Emissaries in Early Modern Literature and Culture*, ed. Charry and Shahani, 61–65.

69. On Roe's discussion of the Mughal copies of the English miniature, see Nandini Das, "'Apes of Imitation': Imitation and Identity in Sir Thomas Roe's Embassy to India," in *A Companion to the Global Renaissance*, ed. Singh, 114–28.

70. As quoted in Beach, "The Mughal Painter Abu'l Hasan and Some English Sources for His Style," 13.

71. On James's marginality in the image, see Peter Stallybrass, "Marginal England: A View from Aleppo," in *Center or Margin*, ed. Orlin and Barroll, 35–36.

72. Natasha Eaton, "Between Mimesis and Alterity: Art, Gift, and Diplomacy in Colonial India, 1770–1800," in *Romantic Representations of British India*, ed. Franklin, 818.

73. On the similarities and differences of hybridity and metamorphosis in the medieval world, some of which resonate here, see Bynum, *Metamorphosis and Identity*, 28–32.

74. On the differences between the temporal natures of hybridity and metamorphosis, see Alison Sharrock, "Representing Metamorphosis," in *Art and Text in Roman Culture*, ed. Elsner, 103–30.

75. Ovid, *Metamorphosis*, 6.63–67. Here I quote Leonard Barkan's translation of the lines as they appear in Barkan, *The Gods Made Flesh*, 3.

76. As evidenced by an entire session devoted to hybridity during the Stone Seminar on Art and Globalization held in Chicago, 2010. See Elkins, Valiavicharska, and Kim, *Art and Globalization*, 51–62.

77. On this problematic paradox of hybridity, see Ballinger, "'Authentic Hybrids' in the Balkan Borderlands," 31–49.

78. A similar theoretical argument has been put forth by Alessandra Russo in an analysis of graffiti scribbled on the walls of the sixteenth-century Augustinian monastery of San Nicolás de Tolentino in the Mexican village of Actopan. In her study, Russo argues that the internal tension of *mestizo* productions "creates mutual transformations among a multiplicity of procedures, techniques, styles, decisions, and outcomes to which it is always possible to add another element of complexity without creating a contradiction," and she, borrowing from Freud, names this transformative and reconcilable aspect of the postconquest *mestizo* images and objects "aesthetic condensation." Russo, "A Tale of Two Bodies," 77.

Bibliography

Abū al-Faẕl ibn Mubārak. *The Akbarnāma of Abu-L-Faẕl*. Bibliotheca Indica, Work no. 138. Calcutta: Asiatic Society, 1897.

Ådahl, Karin, ed. *The Sultan's Procession: The Swedish Embassy to Sultan Mehmed IV in 1657–1658 and the Rålamb Paintings*. Istanbul: Swedish Research Institute in Istanbul, 2006.

Adelson, Howard L. "The Holy Lance and the Hereditary German Monarchy." *Art Bulletin* 48, no. 2 (1966): 177–92.

Agamben, Giorgio. *The Kingdom and the Glory: For a Theological Genealogy of Economy and Government*. Meridian: Crossing Aesthetics. Stanford: Stanford University Press, 2011.

Aikema, Bernard, and Beverly Louise Brown, eds. *Renaissance Venice and the North: Crosscurrents in the Time of Bellini, Dürer, and Titian*. New York: Rizzoli, 2000.

Allgemeine deutsche Biographie. 56 vols. Munich: Duncker und Humboldt, 1875–1912.

Althoff, Gerd. *Spielregeln der Politik im Mittelalter: Kommunikation in Frieden und Fehde*. Darmstadt: Primus, 1997.

Anderson, M. S. *The Rise of Modern Diplomacy, 1450–1919*. London: Longman, 1993.

Appadurai, Arjun, ed. *The Social Life of Things: Commodities in Cultural Perspective*. 1st paperback ed. Cambridge: Cambridge University Press, 1988.

Appuhn, Horst, and Christian von Heusinger. *Riesenholzschnitte und Papiertapeten der Renaissance*. Unterschneidheim: Alfons Uhl, 1976.

Aristotle. *De Anima*. Translated by J. A. Smith. Oxford: Clarendon Press, 1931.

———. *Historia Animalium*, vol. 3. Translated by David M. Balme. Cambridge: Harvard University Press, 1991.

———. *Poetics*. Translated by Stephen Halliwell. Cambridge: Harvard University Press, 1995.

Ashley, Kathleen, and Veronique Plesch. "The Cultural Processes of 'Appropriation.'" *Journal of Medieval and Early Modern Studies* 32, no. 1 (2002): 1–15.

Babcock, Barbara A., ed. *The Reversible World: Symbolic Inversion in Art and Society*. Symbol, Myth, and Ritual. Ithaca: Cornell University Press, 1978.

Bächtold-Stäubli, Hanns, ed. *Handwörterbuch des deutschen Aberglaubens, herausgegeben unter besonderer Mitwirkung*. Handwörterbücher zur deutschen Volkskunde, hrsg. vom Verband deutscher Vereine für Volkskunde. Abt. 1. Aberglaube. Berlin: W. de Gruyter, 1927.

Bacon, Leonard Woolsey, ed. *Dr. Martin Luther's deutsche geistliche Lieder: The Hymns of Martin Luther Set to Their Original Melodies with an English Version*. London: Hoder and Stoughton, 1884.

Bailey, Gauvin A. *Art on the Jesuit Missions in Asia and Latin America, 1542–1773*. Toronto: University of Toronto Press, 1999.

———. "The End of the 'Catholic Era' in Mughal Painting." *Marg* 53 (2001): 46–59.

Bak, János M. *Coronations: Medieval and Early Modern Monarchic Ritual*. Los Angeles: University of California and Center for Medieval and Renaissance Studies, 1985.

Bakhtin, M. M. *Rabelais and His World*. Translated by Hélène Iwolsky. Bloomington: Indiana University Press, 1984.

Ballinger, Pamela. "'Authentic Hybrids' in the Balkan Borderlands." *Current Anthropology* 45, no. 1 (2004): 31–60.

Balmer, R. T. "The Operation of Sand Clocks and Their Medieval Development." *Technology and Culture* 19, no. 4 (1978): 615–32.

Barbour, Richmond. *Before Orientalism: London's Theatre of the East, 1576–1626*. Cambridge Studies in Renaissance Literature and Culture 45. Cambridge: Cambridge University Press, 2003.

Barkan, Leonard. "Diana and Actaeon: The Myth as Synthesis." *English Literary Renaissance* 10, no. 3 (1980): 317–59.

———. *The Gods Made Flesh: Metamorphosis and the Pursuit of Paganism*. New Haven: Yale University Press, 1986.

Barthes, Roland. *Arcimboldo*. Translated by John Shepley. Milan: F. M. Ricci, 1980.

Bath, Michael. *The Image of the Stag: Iconographic Themes in Western Art*. Saecula Spiritalia 24. Baden-Baden: Koerner, 1992.

Baudrillard, Jean. *Simulacra and Simulation*. Ann Arbor: University of Michigan Press, 1994.

Bauer, Rotraud, and Herbert Haupt, eds. "Das Kunstkammerinventar Kaiser Rudolfs II, 1607–1611." *Jahrbuch der Kunsthistorischen Sammlungen in Wien* 72 (1976): 1–191.

Bauernfeind, Hans. *Josef Lieber, Josef Mein*. Blätter für gemischten Chor 238. Vienna: Doblinger, 1960.

Bäumel, Jutta. "'Cave Calvinae—D.N.K.': Das Reichschwert des kursächsischen Kanzlers Dr. Nikolaus Krell von 1601." *Dresdener Kunstblätter* 25, no. 2 (2001): 144–51.

Baumstark, Reinhold, ed. *Das goldene Rössl: Ein Meisterwerk der pariser Hofkunst um 1400*. Munich: Hirmer, 1995.

Beach, Milo Cleveland. "The Gulshan Album and Its European Sources." *Bulletin of the Museum of Fine Arts* 63, no. 332 (January 1, 1965): 63–91.

———. *The Imperial Image: Paintings for the Mughal Court*. Washington, D.C.: Freer Gallery of Art, Smithsonian Institution, 1981.

———. "The Mughal Painter Abu'l Hasan and Some English Sources for His Style." *Journal of the Walters Art Gallery* 38 (1980): 6–33.

Bedini, Silvio A. "The Role of Automata in the History of Technology." *Technology and Culture* 5, no. 1 (1964): 24–42.

Bedos, Brigitte. "The King Enthroned, a New Theme in Anglo-Saxon Royal Iconography: The Seal of Edward the Confessor and Its Political Implications." *Acta / The Center for Medieval and Early Renaissance Studies, State University of New York*, 1986, 53–88.

Beg, Junis, and Alvise Gritti. *Opera nova la quale dechiara tutto il governo del gran turcho*. Venice, 1537.

Belozerskaya, Marina. *Rethinking the Renaissance: Burgundian Arts across Europe*. Cambridge: Cambridge University Press, 2002.

Ben-Amos, Ilana Krausman. "Gifts and Favors: Informal Support in Early Modern England." *Journal of Modern History* 72, no. 2 (2000): 295–338.

Benn, Gottfried. *Briefe*. Wiesbaden: Limes, 1977.

Ben-Zaken, Avner. *Cross-Cultural Scientific Exchanges in the Eastern Mediterranean, 1560–1660*. Baltimore: Johns Hopkins University Press, 2010.

Berns, Jörg Jochen, and Wolfgang Neuber, eds. *Seelenmaschinen: Gattungstraditionen, Funktionen und Leistungsgrenzen der Mnemotechniken vom späten Mittelalter bis zum Beginn der Moderne*. Frühneuzeit-Studien 2. Vienna: Böhlau, 2000.

Berryman, Sylvia. "Ancient Automata and Mechanical Explanation." *Phronesis* 48, no. 4 (2003): 344–69.

Bertele, Hans von, and Erwin Neumann. "Der kaiserliche Kammeruhrmacher Christoph Margraf und die Erfindung der Kugellaufuhr." *Jahrbuch der Kunsthistorischen Sammlungen in Wien* 59 (1963): 39–98.

Bestor, Jane Fair. "Marriage Transactions in Renaissance Italy and Mauss's Essay on the Gift." *Past and Present*, no. 164 (1999): 6–46.

Beutel, Tobias. *Chur=fürstlicher sächsischer stets grünender hoher Cedern=Wald, auf dem grünen Rauten=Grunde, oder Kurtze Vorstellung der chur=fürstl. sächs. hohen Regal=Wercke, nehmlich: Der fürtrefflichen Kunst=Kammer, und anderer, Seiner Chur=Fürstl. Durchl. hochschätzbaren unvergleichlich wichtigen Dinge, allhier bey der Residentz Dreszden . . .* Jn Dreszden, gedruckt bey den Bergischen Erben, anno 1671, 1671.

Bhabha, Homi K. *The Location of Culture*. London: Routledge, 1994.

Bindman, David, Henry Louis Gates, Jr., and Karen C. C. Dalton, eds. *The Image of the Black in Western Art,* Vol. III, Part 1. New ed. Cambridge, Mass.: Belknap Press, 2010.

Bisaha, Nancy. *Creating East and West: Renaissance Humanists and the Ottoman Turks*. Philadelphia: University of Pennsylvania Press, 2004.

Blake, N. F. "English Versions of Reynard the Fox in the Fifteenth and Sixteenth Centuries." *Studies in Philology* 62, no. 1 (1965): 63–77.

Blaschke, Karlheinz. *Moritz von Sachsen: Ein Reformationsfürst der zweiten Generation*. Persönlichkeit und Geschichte 113. Göttingen: Muster-Schmidt, 1983.

Bloch, Maurice. *Ritual, History, and Power: Selected Papers in Anthropology*. Monographs on Social Anthropology 58. London: Athlone Press, 1989.

Boas, Marie. "Hero's Pneumatica: A Study of Its Transmission and Influence." *Isis* 40, no. 1 (1949): 38–48.

Bobinger, Maximilian. "Der augsburger Uhrmacher Hans Schlothaim." *Schriften der Freunde alter Uhren* 11 (1971–72): 4–31.

Bosbach, Franz. "Die politische Bedeutung Karls des Grossen für Karl V." *Archiv für Kulturgeschichte* 84, no. 1 (2002): 113–32.

Bott, Gerhard, ed. *Wenzel Jamnitzer und die nürnberger Goldschmiedekunst, 1500–1700: Goldschmiedearbeiten: Entwürfe, Modelle, Medaillen, Ornamentstiche, Schmuck, Porträts: Eine Ausstellung im Germanischen Nationalmuseum Nürnberg vom 28. Juni–15. September 1985.* Munich: Klinkhardt und Biermann, 1985.

Boyajian, James C. *Portuguese Trade in Asia under the Habsburgs, 1580–1640.* Baltimore: Johns Hopkins University Press, 1993.

Boyd, Kelly. *Encyclopedia of Historians and Historical Writing.* London: Fitzroy Dearborn, 1999.

Brandi, Karl. *Kaiser Karl V: Werden und Schicksal einer Persönlichkeit und eines Weltreiches.* Munich: F. Bruckmann, 1938.

Brant, Sebastian. *Sebastian Brands Niv-Schiff von Narragonien.* Basel: Johann Bergmann, 1494.

Braudel, Fernand. *The Mediterranean and the Mediterranean World in the Age of Philip II.* Vol. 2. Translated by Sian Reynolds. Berkeley: University of California Press, 1996.

Braunfels, Wolfgang. *Die Kunst im Heiligen Römischen Reich Deutscher Nation.* Munich: C. H. Beck, 1979.

Brecht, Martin. *Martin Luther.* Stuttgart: Calwer, 1981.

Bredekamp, Horst. *The Lure of Antiquity and the Cult of the Machine: The Kunstkammer and the Evolution of Nature, Art, and Technology.* Translated by Allison Brown. Princeton: Markus Wiener, 1995.

Brett, Gerard. "The Automata in the Byzantine 'Throne of Solomon.'" *Speculum* 29, no. 3 (1954): 477–87.

Britten, F. J., and G. H. Baillie. *Britten's Old Clocks and Watches and Their Makers: A Historical and Descriptive Account of the Different Styles of Clocks and Watches of the Past in England and Abroad Containing a List of Nearly Fourteen Thousand Makers.* New York: Bonanza, 1956.

Brown, Bill. "Thing Theory." *Critical Inquiry* 28, no. 1 (2001): 1–22.

Bryce, James. *The Holy Roman Empire.* London: Macmillan, 1871.

Bryson, Lyman, ed. *The Communication of Ideas: A Series of Addresses.* Religion and Civilization Series. New York: Harper and Row, 1948.

Bryson, Norman, Michael Ann Holly, and Keith P. F. Moxey, eds. *Visual Culture: Images and Interpretations.* Hanover: University Press of New England for Wesleyan University Press, 1994.

Buc, Philippe. "Ritual and Interpretation: The Early Medieval Case." *Early Medieval Europe* 9, no. 2 (2000): 183–210.

Buettner, Brigitte. "Past Presents: New Year's Gifts at the Valois Courts, ca. 1400." *Art Bulletin* 83, no. 4 (2001): 598–625.

Burke, Jill. "Meaning and Crisis in the Early Sixteenth Century: Interpreting Leonardo's Lion." *Oxford Art Journal* 29, no. 1 (2006): 79–91.

Busbecq, Ogier Ghiselin de. *The Turkish Letters of Ogier Ghiselin de Busbecq, Imperial Ambassador at Constantinople, 1554–1562: Translated from the Latin of the Elzevir Edition of 1663.* Baton Rouge: Louisiana State University Press, 2005.

Butters, Suzanne B. "The Uses and Abuses of Gifts in the World of Ferdinando de' Medici (1549–1609)." *I Tatti Studies in the Italian Renaissance* 11 (2007): 243–354.

Bynum, Caroline Walker. *Metamorphosis and Identity.* New York: Zone Books, 2001.

Cahill, Nicholas. "The Treasury at Persepolis: Gift-Giving at the City of the Persians." *American Journal of Archaeology* 89, no. 3 (1985): 373–89.

Callières, Monsieur de (François). *The Art of Diplomacy.* Leicester: Leicester University Press, 1983.

Camille, Michael. *Image on the Edge: The Margins of Medieval Art.* Essays in Art and Culture. London: Reaktion, 1992.

———. *The Medieval Art of Love: Objects and Subjects of Desire.* London: Laurence King, 1998.

Campbell, Thomas J. *The Jesuits, 1534–1921: A History of the Society of Jesus from Its Foundation to the Present Time.* New York: Encyclopedia Press, 1921.

Campbell, Thomas P. *Tapestry in the Renaissance: Art and Magnificence.* New York: Metropolitan Museum of Art, 2002.

Campori, Giuseppe. *Lettere artistiche inedite.* Modena: Erede Soliani, 1866.

Canby, Sheila R. *Shah 'Abbas: The Remaking of Iran.* London: British Museum Press, 2009.

Carboni, Stefano, ed. *Venice and the Islamic World, 828–1797.* New York: Metropolitan Museum of Art, 2007.

Carroll, Jane Louise, and Alison G. Stewart. *Saints, Sinners, and Sisters: Gender and Northern Art in Medieval and Early Modern Europe.* Aldershot: Ashgate, 2003.

Casel, Odo. *Heilige Überlieferung: Ausschnitte aus der*

Geschichte des Mönchtums und des heiligen Kultes, dem hochwürdigsten Herrn abte von Maria Laach dr. theol. et iur. h. c. Idlefons Herwegen zum silbernen Abtsjubiläume. Beiträge zur Geschichte des alten Mönchtums und des Benediktinerordens. Supplementband. Münster: Aschendorff, 1938.

Casey, Paul F. "Court Performance in Berlin of the Sixteenth Century: Georg Pondo's Christmas Play of 1589." *Daphnis* 32, nos. 1–2 (2003): 57–72.

Certeau, Michel de. *The Mystic Fable.* Translated by Michael B. Smith. Religion and Postmodernism. Chicago: University of Chicago Press, 1992.

Cesareo, Francesco C. "The Collegium Germanicum and the Ignatian Vision of Education." *Sixteenth Century Journal* 24, no. 4 (1993): 829–41.

Chambers, Ephraim. *Cyclopaedia; or, an Universal Dictionary of Arts and Sciences: Containing an Explication of the Terms, and an Account of the Things Signified Thereby, in the Several Arts Both Liberal and Mechanical, and the Several Sciences, Human and Divine. . . .* 4th ed. corr. and amended, with some additions. Goldsmiths' Kress Library of Economic Literature, no. 8048.0 suppl. London: Printed for D. Midwinter, 1741.

Chapuis, Alfred, and Edmond Droz. *Automata: A Historical and Technological Study.* Neuchâtel: Éditions du Griffon, 1958.

Chapuis, Alfred, Edmond Haraucourd, and Édouard Gélis. *Le monde des automates: Étude historique et technique.* Geneva: Slatkine, 1984.

Charry, Brinda, and Gitanjali Shahani. *Emissaries in Early Modern Literature and Culture: Mediation, Transmission, Traffic, 1550–1700.* Farnham: Ashgate, 2009.

Çiğ, Kemal, Sabahattin Batur, and Cengiz Köseoğlu. *The Topkapı Saray Museum.* Translated by J. M. Rogers. Boston: Little, Brown, 1987.

Cipolla, Carlo M. *Clocks and Culture, 1300–1700.* New York: W. W. Norton, 2003.

Clark, William, Jan Golinski, and Simon Schaffer, eds. *The Sciences in Enlightened Europe.* Chicago: University of Chicago Press, 1999.

Clifford, James. *The Predicament of Culture: Twentieth-Century Ethnography, Literature, and Art.* Cambridge: Harvard University Press, 1988.

Cohn, Bernard S. "The Past in the Present: India as Museum of Mankind." *History and Anthropology* 11, no. 1 (1998): 1–38.

Cohn, H. J. "Did Bribes Induce the German Electors to Choose Charles V as Emperor in 1519?" *German History* 19, no. 1 (2001): 1–27.

Colledge, Edmund. *Reynard the Fox and Other Mediaeval Netherlands Secular Literature.* Leiden: A.W. Sijthoff, 1967.

Collins, A. Jefferies, ed. *Jewels and Plate of Queen Elizabeth I: The Inventory of 1574.* London: Trustees of the British Museum, 1955.

Conihout, Isabelle, ed. *Ces curieux navires: Trois automates de la renaissance.* Paris: Réunion des musées nationaux, 2000.

Cooper Union Museum for the Arts of Decoration. *Enamel, an Historic Survey to the Present Day.* New York: Cooper Union Museum for the Arts of Decoration, 1954.

Coster, Will, and Andrew Spicer. *Sacred Space in Early Modern Europe.* Cambridge: Cambridge University Press, 2006.

Crary, Jonathan. *Techniques of the Observer: On Vision and Modernity in the Nineteenth Century.* Cambridge: MIT Press, 1990.

Cunningham, Andrew. *The Four Horsemen of the Apocalypse: Religion, War, Famine, and Death in Reformation Europe.* Cambridge: Cambridge University Press, 2000.

Curipeschitz, Benedict. *Itinerarium der Botschaftsreise des Josef von Lamberg und Niclas Jurischitz durch Bosnien, Serbien, Bulgarien nach Konstantinopel.* Innsbruck: Wagner, 1910.

Cutler, Anthony. "Gifts and Gift Exchange as Aspects of the Byzantine, Arab, and Related Economies." *Dumbarton Oaks Papers* 55 (2001): 247–78.

———. "Significant Gifts: Patterns of Exchange in Late Antique, Byzantine, and Early Islamic Diplomacy." *Journal of Medieval and Early Modern Studies* 38, no. 1 (2008): 79.

Darnton, Robert. "The Symbolic Element in History." *Journal of Modern History* 58, no. 1 (1986): 218–34.

Daston, Lorraine. "Marvelous Facts and Miraculous Evidence in Early Modern Europe." *Critical Inquiry* 18, no. 1 (1991): 93–124.

Dasypodius, Konrad. *Warhafftigte Ausslegung vnd [und] Beschreybung des astronomischen Uhrwercks zu Strassbourg, welches er Anfänglichs Erfunden vnnd [unnd] angeben hat: Auch ein altes Lied vom [sic] dem Kampff und Streyt, so entstanden zwischen dem Roraffen (welcher under der Orglen im Münster zu Strassburg ist) und dem Hanen, so auff der alten Uhren war . . . Der Kampff des Roraffen under der Orglen in Münster zu Strassburg, mit dem Hanen vor 200 Jharen beschriben daselbst, auff dem Uhrwerck.* Strasbourg, 1580.

Davis, Natalie Zemon. *The Gift in Sixteenth-Century France.* The Curti Lectures. Madison: University of Wisconsin Press, 2000.

Dean, Carolyn, and Dana Leibsohn. "Hybridity and Its Discontents: Considering Visual Culture in Colonial Spanish America." *Colonial Latin American Review* 12, no. 1 (2003): 5–35.

Dee, John. *The Mathematicall Praeface to the Elements of Geometrie of Euclid of Megara (1570).* New York: Science History Publications, 1975.

Deri, Max. *Das Rollwerk in der deutschen Ornamentik des sechzehnten und siebzehnten Jahrhunderts.* Berlin: Schuster und Bufleb, 1906.

Deshman, Robert. "Otto III and the Warmund Sacramentary: A Study in Political Theology." *Zeitschrift für Kunstgeschichte* 34, no. 1 (1971): 1–20.

Dewald, Jonathan. *Europe 1450 to 1789: Encyclopedia of the Early Modern World.* New York: Charles Scribner's Sons, 2004.

Dicke, Gerd. *Die Fabeln des Mittelalters und der frühen Neuzeit: Ein Katalog der deutschen Versionen und ihrer lateinischen Entsprechungen.* Münstersche Mittelalter-Schriften 60. Munich: W. Fink, 1987.

Diemer, Dorothea, Peter Diemer, and Lorenz Seelig, eds. *Die münchner Kunstkammer.* Abhandlungen (Bayerische Akademie der Wissenschaften. Philosophisch-Historische Klasse) 129. Munich: Bayerische Akademie der Wissenschaften, 2008.

Diemer, Peter, Elke Bujot, and Dorothea Diemer, eds. *Johannes Baptist Fickler das Inventar der Münchner herzoglichen Kunstkammer von 1598.* Abhandlungen (Bayerische Akademie der Wissenschaften. Philosophisch-Historische Klasse) 125. Munich: Bayerische Akademie der Wissenschaften, 2004.

Doering, Oscar, ed. *Quellenschrifft für Kunstgeschichte und Kunsttechnik des Mittelalter und der Neuzeit.* Vienna: Carl Graser, 1901.

Drachmann, A. G. *Ktesibios, Philon and Heron: A Study in Ancient Pneumatics.* Copenhagen: Munksgaard, 1948.

———. *The Mechanical Technology of Greek and Roman Antiquity: A Study of the Literary Sources.* Copenhagen: Munksgaard, 1963.

Dückers, Rob. *The Limbourg Brothers: Nijmegen Masters at the French Court, 1400–1416.* Nijmegen: Ludion, 2005.

Dürer, Albrecht. *Albrecht Dürer: Das druckgraphische Werk.* Munich: Prestel, 2004.

Eichberger, Dagmar, and Yvonne Bleyerveld. *Women of Distinction: Margaret of York, Margaret of Austria.* Leuven: Davidsfonds, 2005.

Eichberger, Dagmar, and Charles Zika, eds. *Dürer and His Culture.* Cambridge: Cambridge University Press, 1998.

Eisler, Colin. "Two Early Franco-Flemish Embroideries: Suggestions for Their Settings." *Burlington Magazine* 109, no. 775 (1967): 571–81.

Elkins, James, Zhivka Valiavicharska, and Alice Kim. *Art and Globalization.* Stone Art Theory Institutes. University Park: Pennsylvania State University Press, 2010.

Elsner, Jaś. *Art and Text in Roman Culture.* Cambridge Studies in New Art History and Criticism. Cambridge: Cambridge University Press, 1996.

———. *Roman Eyes: Visuality and Subjectivity in Art and Text.* Princeton: Princeton University Press, 2007.

Engelmann, Max. "Das Krippenwerk des Augsburgers Hans Schlottheim." *Kunstwanderer* 3 (1921): 147–50.

Esprinchard, Jacques. *Vie de Jacques Esprinchard, Rochelais, et journal de ses voyages au XVIe siècle.* Bibliothèque générale de l'École pratique des hautes études, sciences économiques et sociales. Paris: SEVPEN, 1957.

Ettinghausen, Richard. *Paintings of the Sultans and Emperors of India in American Collections.* Lalit Kalā Series of Indian Art. New Delhi: Lalit Kalā Akademi, 1961.

Evans, Jessica, and Stuart Hall. *Visual Culture: The Reader.* London: SAGE, 1999.

Evans, Robert John Weston. *Rudolf II and His World: A Study in Intellectual History, 1576–1612.* Oxford: Clarendon Press, 1984.

Evans, Robert John Weston, and Alexander Marr. *Curiosity and Wonder from the Renaissance to the Enlightenment.* Burlington: Ashgate, 2006.

Farago, Claire J., ed. *Reframing the Renaissance: Visual Culture in Europe and Latin America, 1450–1650.* New Haven: Yale University Press, 1995.

Fazl, Abu. *Akbarnama.* Translated by H. Beveridge. Calcutta: Asiatic Society, 1897–1913.

Fernández Álvarez, Manuel. *Charles V: Elected Emperor and Hereditary Ruler.* London: Thames and Hudson, 1975.

Ferry, Patrick T. "Confessionalization and Popular Preaching: Sermons against Synergism in Reformation Saxony." *Sixteenth Century Journal* 28, no. 4 (1997): 1143–66.

Fichtner, Paula Sutter. *Terror and Tolerance: The Habsburg Empire Confronts Islam, 1526–1850.* London: Reaktion, 2008.

Flood, Barry, David Joselit, Alexander Nagel, Alessandra Russo, Eugene Wang, Christopher Wood, and Mimi Yiengpruksawan. "Roundtable: The Global before Globalization." *October* 2010, 3–19.

Flood, Finbarr Barry. *Objects of Translation: Material Culture and Medieval "Hindu-Muslim" Encounter.* Princeton: Princeton University Press, 2009.

Flores, Jorge Manuel, ed. *Goa and the Great Mughal.* Lisbon: Calouste Gulbenkian Foundation, 2004.

Forssman, Erik. "Die Prunkkassette mit der Allegorie der Philosophie von Wenzel Jamnitzer." *Dresdener Kunstblätter* 48 no. 4 (2004): 246–49.

———. *Wenzel Jamnitzer und die nürnberger Goldschmiedekunst, 1500–1700.* Munich: Klinkhardt und Biermann, 1985.

Frank, Andre Gunder. *ReOrient: Global Economy in the Asian Age.* Berkeley: University of California Press, 1998.

Frankl, Paul. "Sustris und die Münchner Michaelskirche." *Münchner Jahrbuch der bildenden Kunst* 10 (1916): 141–51.

Franklin, Michael J. *Romantic Representations of British India.* Routledge Studies in Romanticism 7. London: Routledge, 2006.

Fraser, J. T., and Nathaniel Lawrence. *The Study of Time II: Proceedings of the Second Conference of the International Society for the Study of Time, Lake Yamanaka, Japan.* New York: Springer, 1975.

Freedberg, David. *The Power of Images: Studies in the History and Theory of Response.* Chicago: University of Chicago Press, 1989.

Fučíková, Eliška, et al. *Rudolf II and Prague: The Court and the City.* Prague: Prague Castle Administration, 1997.

Fudge, Erica. *Brutal Reasoning: Animals, Rationality, and Humanity in Early Modern England.* Ithaca: Cornell University Press, 2006.

Füssel, Stephan, ed. *Biblia: Das ist die gantze heilige Schrifft.* Bible de Luther de 1534: Réimpression intégrale. Cologne: Taschen, 2003.

Gamboni, Dario, Georg Germann, and François de Capitani. *Zeichen der Freiheit: Das Bild der Republik in der Kunst des 16. bis 20. Jahrhunderts.* Bern: Stämpfli, 1991.

García-Diego, José A. *Juanelo Turriano, Charles V's Clockmaker: The Man and His Legend.* Translated by Charles David Ley. Wadhurst: Antiquarian Horological Society, 1986.

Geertz, Clifford. *Negara: The Theatre State in Nineteenth-Century Bali.* Princeton: Princeton University Press, 1980.

Geiger, Johann. *Horologium politicum: Das ist: einfältiger Discurs und kurtze Betrachtung dess politischen, materialischen, selbst gehenden unnd schlagenden Uhrwercks; desselben Nutz und Gebrauch, zu sampt andern Automatibus und selbst bewegenden Wercken . . . ; In acht Capitel verfasst. . . .* Nuremberg, 1621.

Gell, Alfred. *Art and Agency: An Anthropological Theory.* Oxford: Clarendon Press, 1998.

———. "The Art of Anthropology: Essays and Diagrams." London: Athlone Press, 1999.

Gennep, Arnold van. *The Rites of Passage.* Chicago: University of Chicago Press, 1960.

Gerlach, Stefan, Samuel Gerlach, Tobias Wagner, and Heinrich Friese. *Stephan Gerlachs dess aeltern Tage-Buch: der von zween glorwürdigsten Römischen Käysern, Maximiliano und Rudolpho, beyderseits den Andern dieses Nahmens, höchstseeligster Gedächtnüss* [sic], *an die Ottomannische Pforte zu Constantinopel abgefertigten . . . und durch . . . David Ungnad . . . mit würcklicher Erhalt und Verlängerung dess Friedens, zwischen dem Ottomannischen und Römischen Käyserthum und demselben angehörigen Landen und Königreichen &c. glücklichst-vollbrachter Gesandtschafft.* Frankfurt am Main, 1674.

Geuffroy, Antoine. *Briesve description de la court du Grant Turc, et ung sommaire du règne des Ottmans. Avec ung abregé de leurs folles superstitions, ensemble l'origine de cinq empires . . . de la secte de Mehemet.* Paris: Chrestien Wechel, 1546.

———. *Estat de la court du Grant Turc, l'ordre de sa gendarmerie, et ses finances, avec ung brief discours de leurs conquestes depuis le premier de ceste race / [par A. Geuffroy].* Paris: Chrestien Wechel, 1542.

Gévay, Anton von, ed. *Urkunden und Actenstücke zur Geschichte der Verhältnisse zwischen Österreich, Ungern und der Pforte im XVI. und XVII. Jahrhunderte aus Archiven und Bibliotheken: Gesandtschaft König Ferdinands I. an Sultan Suleiman I., 1532–1533.* Vienna: Auf Kosten des Herausgebers, 1838.

Glaser, Hubert. *Beiträge zur bayerischen Geschichte und Kunst, 1573–1657.* Munich: Hirmer, 1980.

Goetz, Hans-Werner. *Moderne Mediävistik: Stand und Perspektiven der Mittelalterforschung.* Darmstadt: Primus, 1999.

Goetz, Walter. *Die bayerische Politik im ersten Jahrzehnt der Regierung Herzog Albrechts V. von Baiern (1550–1560).* Munich: M. Rieger, 1895.

Goffman, Daniel. *The Ottoman Empire and Early Modern Europe.* New Approaches to European History 24. Cambridge: Cambridge University Press, 2002.

Gombrich, E. H. *Art and Illusion: A Study in the Psychology of Pictorial Representation.* Princeton: Princeton University Press, 2000.

———. "Moment and Movement in Art." *Journal of the Warburg and Courtauld Institutes* 27 (1964): 293–306.

Gotthard, Axel. "'Politice seint wir bäpstisch': Kursachsen und der deutsche Protestantismus im frühen 17. Jahrhundert." *Zeitschrift für historische Forschunge* 20 (1993): 275–319.

Grassi, Ernesto, and Maristella Lorch. *Folly and Insanity in Renaissance Literature*. Binghamton: Medieval and Renaissance Texts and Studies, 1986.

Grimm, Jacob. *Deutsches Wörterbuch . . . Quellenverzeichnis*. Leipzig: Hirzel, 1966.

Groebner, Valentin. *Liquid Assets, Dangerous Gifts: Presents and Politics at the End of the Middle Ages*. Philadelphia: University of Pennsylvania Press, 2002.

Grössinger, Christa. *The World Upside-Down: English Misericords*. London: Harvey Miller, 1997.

Gruber, Alain Charles, and Bruno Pons. *L'Art décoratif en Europe*. Paris: Citadelles et Mazenod, 1992.

Gruber, Alain Charles, and Margherita Azzi Visentini. *The History of Decorative Arts*. New York: Abbeville Press, 1994.

Gruzinski, Serge. *Images at War: Mexico from Columbus to "Blade Runner" (1492–2019)*. Durham: Duke University Press, 2001.

Guerreiro, Fernão. *Jahangir and the Jesuits, with an Account of the Travels of Benedict Goes and the Mission to Pegu, from the Relations of Father Fernão Guerreiro, S.J. . . .* London: G. Routledge and Sons, 1930.

Gutfleisch, Barbara, and Joachim Menzhausen. "'How a Kunstkammer Should Be Formed': Gabriel Kaltemarckt's Advice to Christian I of Saxony on the Formation of an Art Collection, 1587." *Journal of the History of Collections* 1, no. 1 (1989): 3–32.

Haas, Alois M. *Sermo mysticus: Studien zu Theologie und Sprache der deutschen Mystik*, vol. 4. Freiburg: Universitätsverlag, 1979.

Haase, Gisela, and Jürgen Karpinski. *Sächsisches Glas: Geschichte, Zentren, Dekorationen*. Munich: Klinkhardt und Biermann, 1988.

Häberlein, Mark, and Johannes Burkhardt. *Die Welser: Neue Forschungen zur Geschichte und Kultur des oberdeutschen Handelshauses*. Berlin: Akademie, 2002.

Habsburg, Francesca von. *The St. Petersburg Muraqqa': Album of Indian and Persian Miniatures from the 16th through the 18th Century and Specimens of Persian Calligraphy by 'Imād Al-Ḥasanī*. Milan: Leonardo Arte, 1996.

Hainhofer, Philipp. "Des Augsburger Patriciers Philipp Hainhofer Reisen nach Innsbruck und Dresden." In *Quellenschrifft für Kunstgeschichte und Kunsttechnik des Mittelalter und der Neuzeit*, edited by Rudolf Eitelberger von Edelberg. Vienna: Carl Graeser, 1901.

Hamburger, Jeffrey F. *St. John the Divine: The Deified Evangelist in Medieval Art and Theology*. Berkeley: University of California Press, 2002.

Hamm, Berndt. "Frömmigkeit als Gegenstand theologiegeschichtlicher Forschung: Methodisch-Historische Überlegungen am Beispiel von Spätmittelalter und Reformation." *Zeitschrift für Theologie und Kirche* 74, no. 4 (1977): 464–97.

———. "Normative Centering in the Fifteenth and Sixteenth Centuries: Observations on Religiosity, Theology, and Iconology." *Journal of Early Modern History* 3, no. 3 (1999): 307–54.

Harper, James G., ed. *The Turk and Islam in the Western Eye, 1450–1750: Visual Imagery before Orientalism*. Transculturalisms, 1400–1700. Burlington: Ashgate, 2011.

Haspels, J. J. L., ed. *Royal Music Machines*. Utrecht: Nationaal Museum van Speelklok tot Pierement, 2006.

Hassig, Debra, ed. *The Mark of the Beast: The Medieval Bestiary in Art, Life, and Literature*. New York: Garland, 1999.

———. *Medieval Bestiaries: Text, Image, Ideology*. Cambridge: Cambridge University Press, 1995.

Hausberger, Isolde, and Rolf Biedermann, eds. *Hans Burgkmair 1473–1973: Das graphische Werk*. Stuttgart: Graphische Sammlung, Staatsgalerie, 1973.

Häutle, Christian. "Die Reisen des Augsburgers Philipp Hainhofer nach Eichstädt, München und Regensburg in den Jahren 1611, 1612 und 1613." *Zeitschrift des historischen Vereins für Schwaben und Neuburg* 8 (1881).

Hayward, John F. "The Mannerist Goldsmiths, 5: Northern Germany, Part VII." *Connoisseur* 175 (1970): 22–30.

———. *Virtuoso Goldsmiths and the Triumph of Mannerism, 1540–1620*. London: Sotheby Parke Bernet, 1976.

Heer, Friedrich. *The Holy Roman Empire*. Translated by Janet Sondheimer. London: Weidenfeld and Nicolson, 1968.

Heigel, Karl Theodor. "Die Censur in Altbaiern." *Archiv für Geschichte des deutschen Buchhandels* 2 (1879): 5–32.

Hentschel, Walter. "Epitaphe in der alten dresdner Frauenkirche: Eine Untersuchung nach Zeichnungen im Kupferstich-Kabinett der Staatlichen Kunstsammlungen Dresden." *Jahrburch Staatliche Kunstsammlungen Dresden, 1961–62*, 101–25.

Herwig, Wolfram, ed. *Österreichische Geschichte*. Vienna: Ueberreuter, 1994–2003.

Hillerbrand, Hans Joachim. *The Oxford Encyclopedia of the Reformation*. New York: Oxford University Press, 1996.

Hilsdale, Cecily J. "Constructing a Byzantine 'Augusta': A Greek Book for a French Bride." *Art Bulletin* 87, no. 3 (2005): 458–83.

Hind, Arthur Mayger. *Early Italian Engraving: A Critical Catalogue with Complete Reproduction of All the Prints Described.* London: Published for M. Knoedler, New York, by B. Quaritch, 1938–48.

Hirschbiegel, Jan, and Werner Paravicini, eds., *Das Frauenzimmer: Die Frau bei Hofe in Spätmittelalter und früher Neuzeit: 6. Symposium der Residenzen-Kommission der Akademie der Wissenschaften in Göttingen, veranstaltet in Zusammenarbeit mit dem Deutschen Historischen Institut Paris, dem Sonderforschungsbereich 537 der Technischen Universität Dresden und dem Landsamt für Archäologie des Freistaates Sachsen, Dresden, 26. bis 29. September 1998.* Residenzenforschung, Bd. 11. Stuttgart: J. Thorbecke, 2000.

Hitchcock, Henry Russell. *German Renaissance Architecture.* Princeton: Princeton University Press, 1981.

Hoberg, Hermann. *Die Inventare des päpstlichen Schatzes in Avignon, 1314–1376.* Studi e testi (Biblioteca apostolica vaticana) 111. Vatican City: Biblioteca apostolica vaticana, 1944.

Hoffman, Eva Rose F. *Late Antique and Medieval Art of the Mediterranean World.* Malden, Mass.: Blackwell, 2007.

Holdenried, Anke. *The Sibyl and Her Scribes: Manuscripts and Interpretation of the Latin Sibylla Tiburtina c. 1050–1500.* Burlington: Ashgate, 2006.

Holzhausen, Walter, and Erna V. Watzdorf. "Brandenburgisch-Sächsische Wachsplastik des XVI. Jahrhunderts: Studien aus den Kunstkammern in Berlin und Dresden," *Jahrbuch der Preuszischen Kunstsammlungen* 52 (1931): 235–257.

Homer. *The Iliad.* Translated by Robert Fitzgerald. Oxford: Oxford University Press, 2008.

Honegger, Thomas. "'A Fox Is a Fox Is a Fox': *The Fox and the Wolf Reconsidered.*" *Reinardus* 9 (1996): 59–74.

Houben, Jan E. M. *Ideology and Status of Sanskrit: Contributions to the History of the Sanskrit Language.* Brill's Indological Library, vol. 13. Leiden: E. J. Brill, 1996.

Hoyer, G., and H. Stierhof, eds. *Schatzkammer der Residenz München: Amtlicher Führer.* Munich: Bayerische Verwaltung der Staatlichen Schlösser, Gärten und Seen, 1976.

Hsia, R. Po-chia. *Social Discipline in the Reformation: Central Europe, 1550–1750.* New York: Routledge, 1989.

Hupp, Otto. *Kunstschätze des regensburger Rathauses.* Regensburg: Habbel, 1912.

Hyman, Wendy Beth, ed. *The Automaton in English Renaissance Literature.* Literary and Scientific Cultures of Early Modernity. Farnham: Ashgate, 2011.

Imber, Colin, and Keiko Kiyotaki. *Frontiers of Ottoman Studies: State, Province, and the West.* Library of Ottoman Studies. New York: I. B. Tauris, 2005.

İnalcık, Halil. *The Ottoman Empire: The Classical Age, 1300–1600.* London: Phoenix, 1994.

İnalcık, Halil, and Cemal Kafadar. *Süleymân the Second and His Time.* Istanbul: Isis Press, 1993.

İnalcık, Halil, and Donald Quataert, eds. *An Economic and Social History of the Ottoman Empire, 1300–1914.* Cambridge: Cambridge University Press, 1994.

International Congress of the History of Art. *L'art et les révolutions: Actes [du] XXVIIe congrès international d'histoire de l'art, Strasbourg, 1–7 septembre 1989.* Strasbourg: Société alsacienne pour le développement de l'histoire de l'art, 1992.

———. *Crossing Cultures: Conflict, Migration and Convergence: The Proceedings of the 32nd International Congress of the History of Art.* Carlton: Miegunyah Press, Melbourne University Publishing, 2009.

Ionus, and Luigi Gritti. *Opera noua la quale dechiara tutto il governo del Gran Turcho: & tutta la spesa che il Gran Turcho ha sotto di lui cosi in pace como in guerra, & il numero de le persone & nome & governo de le sue donnne & garzoni che lui tent nel serraglio serrati & de tutta la entrata che lui ha a lanno & nomina tutti li signori de le sue provincie. . . .* Venice, 1537.

Isenburg, Wilhelm Karl. *Stammtafeln zur Geschichte der europäischen Staaten: Europäische Stammtafeln.* Marburg: Stargardt, 1960.

Isidorus of Seville, Saint. *The Etymologies of Isidore of Seville.* Translated by Stephen A. Barney et al. Cambridge: Cambridge University Press, 2011.

Jahangir. *The Jahangirnama: Memoirs of Jahangir, Emperor of India.* Washington, D.C.: Freer Gallery of Art, Arthur M. Sackler Gallery, Smithsonian Institution, 1999.

Jahrbuch der Kunsthistorischen Sammlungen des Allerhöchsten Kaiserhauses. Vienna: A. Schroll, 1883–1918.

Janson, H. W. *Apes and Ape Lore in the Middle Ages and the Renaissance.* Studies of the Warburg Institute 20. London: Warburg Institute, University of London, 1952.

Jardine, Lisa. *Worldly Goods: A New History of the Renaissance.* London: W. W. Norton, 1996.

Jardine, Lisa, and Jerry Brotton. *Global Interests: Renaissance Art between East and West.* Ithaca: Cornell University Press, 2000.

Jazarī, Ismā'īl ibn al-Razzāz. *The Book of Knowledge of*

Ingenious Mechanical Devices. Translated by Donald R. Hill. Dordrecht: D. Reidel, 1974.

Johnson, Carina L. *Cultural Hierarchy in Sixteenth-Century Europe: The Ottomans and Mexicans.* Cambridge: Cambridge University Press, 2011.

Jonassen, Frederick B. "Lucian's 'Saturnalia,' the Land of Cockaigne, and the Mummers' Plays." *Folklore* 101, no. 1 (1990): 58–68.

Kafadar, Cemal. *Between Two Worlds: The Construction of the Ottoman State.* Berkeley: University of California Press, 1995.

Kang, Minsoo. *Sublime Dreams of Living Machines: The Automaton in the European Imagination.* Cambridge: Harvard University Press, 2010.

Kantorowicz, Ernst Hartwig. *The King's Two Bodies: A Study in Mediaeval Political Theology.* Princeton: Princeton University Press, 1997.

Kaplan, Paul H. D. *The Rise of the Black Magus in Western Art.* Ann Arbor, Mich.: UMI Research Press, 1985.

Kaufman, Peter Iver. "Luther's 'Scholastic Phase' Revisited: Grace, Works, and Merit in the Earliest Extant Sermons." *Church History* 51, no. 3 (1982): 280–89.

Kaufmann, Thomas DaCosta. *Arcimboldo: Visual Jokes, Natural History, and Still-Life Painting.* Chicago: University of Chicago Press, 2009.

———. *The Mastery of Nature: Aspects of Art, Science, and Humanism in the Renaissance.* Princeton: Princeton University Press, 1993.

———. "Remarks on the Collections of Rudolf II: The Kunstkammer as a Form of Representation." *Art Journal* 38, no. 1 (1978): 22–28.

———. *The School of Prague: Painting at the Court of Rudolf II.* Chicago: University of Chicago Press, 1988.

———. *Toward a Geography of Art.* Chicago: University of Chicago Press, 2004.

———. *Variations on the Imperial Theme in the Age of Maximilian II and Rudolf II.* Outstanding Dissertations in the Fine Arts. New York: Garland, 1978.

Keating, Jessica. "The Machinations of German Court Culture: Early Modern Automata." PhD diss., Northwestern University, 2010.

Kellenbenz, Hermann. *Die Fugger in Spanien und Portugal bis 1560: Ein Grossunternehmen des 16. Jahrhunderts.* Studien zur Fuggergeschichte 32. Munich: Ernst Vögel, 1990.

Kemp, Martin. "Navis Ecclesiae: An Ambrosian Metaphor in Leonardo's Allegory of the Nautical Wolf and Imperious Eagle." *Bibliothèque d'humanisme et renaissance* 43, no. 2 (1981): 257–68.

———. "Taking It on Trust: Form and Meaning in Naturalistic Representation." *Archives of Natural History* 17, no. 2 (1990): 127–88.

Kivistö, Sari. *The Vices of Learning: Morality and Knowledge at Early Modern Universities.* Boston: Brill, 2014.

Klarwill, Victor, ed. *Fugger-Zeitungen: Ungedruckte Briefe an das Haus Fugger aus den Jahren 1568–1605.* Vienna: Rikola, 1923.

Koch, Ebba. *Mughal Art and Imperial Ideology: Collected Essays.* New York: Oxford University Press, 2001.

Koerner, Joseph Leo. *The Moment of Self-Portraiture in German Renaissance Art.* Chicago: University of Chicago Press, 1993.

———. "The Mortification of the Image: Death as a Hermeneutic in Hans Baldung Grien." *Representations*, no. 10 (1985): 52–101.

———. *The Reformation of the Image.* London: Reaktion, 2004.

Kohler, Alfred. *Karl V. 1500–1558: Eine Biographie.* Munich: C. H. Beck, 1999.

Kohlhaussen, Heinrich. *Minnekästchen im Mittelalter.* Berlin: Verlag für Kunstwissenschaft, 1928.

Kołakowski, Leszek. *God Owes Us Nothing: A Brief Remark on Pascal's Religion and on the Spirit of Jansenism.* Chicago: University of Chicago Press, 1995.

König, Maria Angela. *Weihegaben an U. L. Frau von Altötting, vom Beginn der Wallfahrt bis zum Abschluss der Säkularisation.* Munich: Lentner, 1939.

Königlich Bayerische Akademie der Wissenschaften, Philosophisch-Philologische Klasse, Königlich Bayerische Akademie der Wissenschaften, and Historische Klasse. *Sitzungsberichte der philosophisch-philologischen und historischen Klasse der K. B. Akademie der Wissenschaften zu München*, 1871.

Koziol, Geoffrey. *Begging Pardon and Favor: Ritual and Political Order in Early Medieval France.* Ithaca: Cornell University Press, 1992.

Kretzenbacher, Leopold. *Frühbarockes Weihnachtsspiel in Kärnten und Steiermark: Klagenfurter und Grazer Weihnachtsspieltexte des frühen 17. Jahrhunderts als kulturhistorische Denkmäler der Gegenreformation in Innerösterreich.* Archiv für vaterländische Geschichte und Topographie 40. Klagenfurt: Geschichtsverein für Kärnten, 1952.

Kruijtzer, Gijs. *Xenophobia in Seventeenth-Century India.* LUP Dissertations. Amsterdam: Leiden University Press, 2009.

Kubler, George. *The Shape of Time: Remarks on the History of Things.* New Haven: Yale University Press, 1962.

Kugler, Georg Johannes, and Wilfried Seipel. *Kaiser Ferdinand I. 1503–1564: Das Werden der Habsburgermonarchie: Kunsthistorisches Museum*. Vienna: Kunsthistorisches Museum, 2003.

Kunsthistorisches Museum Wien, Sammlung für Plastik und Kunstgewerbe. *Katalog der Sammlung für Plastik und Kunstgewerbe*. Vol. 1, *Mittelalter*. Vienna: Kunsthistorisches Museum Wien, 1964.

Kunsthistorisches Museum (Sammlungen Schloss). *Die Kunstkammer*. Innsbruck: 1977.

Kuripešić, Benedikt. *Itinerarium, oder: Wegrayss küniglich Mayestät potschafft gen Constantinopel zudem türckischen Keiser Soleyman, Anno 1530*. Österreichisch-bosnische Beziehungen 2. Klagenfurt: Wieser, 1997.

Kurz, Otto. *The Decorative Arts of Europe and the Islamic East: Selected Studies*. London: Dorian Press, 1977.

———. *European Clocks and Watches in the Near East*. Studies of the Warburg Institute 34. London: Warburg Institute, University of London, 1975.

Labarte, Jules, ed. *Inventaire du mobilier de Charles V, roi de France*. Collection de documents inédits sur l'histoire de France, troisième série, archéologie. Paris: Imprimerie nationale, 1879.

Lach, Donald F. *Asia in the Making of Europe*. Chicago: University of Chicago Press, 1993.

LaGrandeur, Kevin. *Androids and Intelligent Networks in Early Modern Literature and Culture: Artificial Slaves*. New York: Routledge, 2013.

Lamalle, Edmond, and Nicolas Trigault. *La propagande du P. Nicolas Trigault: En faveur des missions de Chine (1616)*. Archivum Historicum Societatis Jesu 9, no. 1. Roma: Spirito, 1940.

Lamb, John David. *Josef lieber, Josef mein*. Berkeley: Wymm, 1965.

Lawrence, Christopher, and Steven Shapin. *Science Incarnate: Historical Embodiments of Natural Knowledge*. Chicago: University of Chicago Press, 1998.

Leitch, Stephaine. "Burgkmair's *Peoples of Africa and India* (1508) and the Origins of Ethnography in Print." *Art Bulletin* 91, no. 2 (2009): 134–59.

———. *Mapping Ethnography in Early Modern Germany: New Worlds in Print Culture*. New York: Palgrave Macmillan, 2010.

Levenson, Jay A., ed. *Circa 1492: Art in the Age of Exploration*. Washington, D.C.: National Gallery of Art, 1991.

Levenson, J., K. Oberhuber, and J. Sheehan. *Early Italian Engravings from the National Gallery of Art*. Washington, D.C.: National Gallery of Art, 1973.

Leyser, Karl. *Communications and Power in Medieval Europe: The Carolingian and Ottonian Centuries*. London: Hambledon Press, 1994.

Lietzmann, Hilda. "Die Geschichte zweier Automaten: Ein weiterer Beitrag zum Werk des Valentin Drausch." *Zeitschrift für Kunstgeschichte* 57, no. 3 (1994): 390–402.

———. "Der kaiserliche Antiquar Jacopo Strada und Kurfürst August von Sachsen." *Zeitschrift für Kunstgeschichte* 60, no. 3 (1997): 377–99.

———. *Valentin Drausch und Herzog Wilhelm V. von Bayern: Ein Edelsteinschneider der Spätrenaissance und sein Auftraggeber*. Munich: Deutscher Kunstverlag, 1998.

Lightbown, R.W. *French Silver*. London: H. M. Stationery Office, 1978.

Littlefield, Sharon. "The Object in the Gift: Embassies of Jahangir and Shah Abbas." PhD diss., University of Minnesota, 1999.

Liudprand of Cremona. *The Embassy to Constantinople and Other Writings*. London: Dent, 1993.

Lockhart, Paul Douglas. *Frederik II and the Protestant Cause: Denmark's Role in the Wars of Religion, 1559–1596*. Leiden: E. J. Brill, 2004.

Loebl, Alfred H. *Zur Geschichte des Türkenkrieges von 1593–1606*. Prague: Praguer Studien aus dem Gebiet der Geschichtswissenschaft, 1899.

Lomazzo, Giovanni Paolo. *Scritti sulle arti*. Edited by Roberto Paolo Ciardi. Raccolta pisana di saggi e studi, nos. 33, 34. Florence: Marchi e Bertolli, 1973–74.

Louth, Andrew. "Augustine on Language." *Literature and Theology* 3, no. 2 (1989): 151–58.

Luther, Martin. *D. Martin Luthers Werke: Kritische Gesamtausgabe*. Weimar: H. Böhlaus, 1883–2009.

———. *Dr. Martin Luthers deutsche geistliche Lieder*. Breitkopf und Härtel, 1840.

———. *Sermons of Martin Luther*. Grand Rapids, Mich.: Baker Book House, 1983.

———. *Works of Martin Luther: With Introductions and Notes*. Edited by Henry Eyster Jacobs and Adolph Spaeth. Philadelphia: Muhlenberg Press, 1943.

Macey, Samuel L., ed. *Encyclopedia of Time*. New York: Garland, 1994.

Maclagan, Edward. *The Jesuits and the Great Mogul*. London: Burns, Oates, and Washbourne, 1932.

Majoros, Ferenc. *Karl V.: Habsburg als Weltmacht*. Graz: Styria, 2000.

Malebranche, Nicolas. *Treatise on Nature and Grace*. Translated by Patrick Riley. Oxford: Clarendon Press, 1992.

Malekandathil, Pius. *The Germans, the Portuguese, and India*. Münster: LIT, 1999.

Malinowski, Bronislaw. *Argonauts of the Western Pacific: An Account of Native Enterprise and Adventure in the Archipelagoes of Melanesian New Guinea.* London: Routledge and Sons, 1922.

Marx, Barbara. *Kunst und Repräsentation am dresdner Hof.* Munich: Deutscher Kunstverlag, 2005.

Matar, N. I. *Turks, Moors, and Englishmen in the Age of Discovery.* New York: Columbia University Press, 1999.

Matheson, Peter. *Argula von Grumbach: A Woman's Voice in the Reformation.* Edinburgh: T and T Clark, 1995.

Mathew, Kuzhippalli Skaria. *Indo-Portuguese Trade and the Fuggers of Germany: Sixteenth Century.* New Delhi: Manohar, 1997.

Mathews, Thomas F. *The Clash of Gods: A Reinterpretation of Early Christian Art.* Princeton: Princeton University Press, 1993.

Mathias, Peter. "Skills and the Diffusion of Innovations from Britain in the Eighteenth Century." *Transactions of the Royal Historical Society* 25 (1975): 93–113.

Mattingly, Garrett. *Renaissance Diplomacy.* Boston: Houghton Mifflin, 1971.

Maurice, Klaus. *Die deutsche Räderuhr: Zur Kunst und Technik des mechanischen Zeitmessers im deutschen Sprachraum.* Munich: C. H. Beck, 1976.

Maurice, Klaus, and Otto Mayr, eds. *The Clockwork Universe: German Clocks and Automata, 1550–1650.* Washington, D.C.: Smithsonian Institution, 1980.

Mauss, Marcel, and E. E. Evans-Pritchard. *The Gift: Forms and Functions of Exchange in Archaic Societies.* New York: W. W. Norton, 1967.

Maxwell, Susan. *The Court Art of Friedrich Sustris: Patronage in Late Renaissance Bavaria.* Burlington: Ashgate, 2011.

Mayr, Otto. *Authority, Liberty, and Automatic Machinery in Early Modern Europe.* Baltimore: Johns Hopkins University Press, 1986.

McDonald, Mark P. "Burgkmair's Woodcut Frieze of the Natives of Africa and India." *Print Quarterly* 20, no. 3 (2003): 227–44.

McGinn, Bernard. *Visions of the End: Apocalyptic Traditions in the Middle Ages.* New York: Columbia University Press, 1998.

Meiss, Millard, ed. *De Artibus Opuscula XL: Essays in Honor of Erwin Panofsky.* New York: New York University Press, 1961.

Mellinkoff, Ruth. *Outcasts: Signs of Otherness in Northern European Art of the Late Middle Ages.* Berkeley: University of California Press, 1993.

Menzhausen, Joachim. "Kurfürst Augusts Kunstkammer: Eine Analyse des Inventars von 1587." *Jahrbuch der staatlichen Kunstsammlungen Dresden* 17 (1985): 21–29.

———. "Kürfurst Augusts Kunstkammer"; Dirk Syndram, *Die Inventare der kurfürstlich-sächsischen Kunstkammer in Dresden: 1587* (Dresden: Sandstein, 2010).

Meserve, Margaret. *Empires of Islam in Renaissance Historical Thought.* Cambridge: Harvard University Press, 2008.

Meyers kleines Konversations-Lexikon: Fünfte, umgearbeitete und vermehrte Auflage. Mit uber 100 Beilagen, Karten und Bildertafeln . . . Leipzig, 1893.

Michaelis, Johann Gottfried. *Dressdnische Inscriptiones und Epitaphia: Welche auf denen Monumentis derer in Gott ruhenden, so allhier in und ausser der Kirche zu unser lieben Frauen begraben liegen, und eine fröliche Aufferstehung erwarten, zu finden . . .* Alt-Dressden: Schwencke, 1714.

———. *Johann Gottfried Michaelis Monumenta Dresdensia: Oder Grab u. Ehren-Mahle d. Haupt-Stadt u. Vestung Dresden; besonders derer daselbst in Gott ruhenden, in u. ausser d. Kirchen zur lieben Frauen . . .* Budisin: Richter, 1718.

Michalski, Sergiusz. *The Reformation and the Visual Arts: The Protestant Image Question in Western and Eastern Europe.* New York: Routledge, 1993.

Migne, Jacques-Paul, ed. *Patrologiae cursus completus: Series graeca.* Paris, 1857–66.

———. *Patrologiae cursus completus: Series latina.* Paris, 1844–55.

Mitrovic, Václav Vratislav z. *Adventures of Baron Wenceslas Wratislaw of Mitrowitz: What He Saw in the Turkish Metropolis, Constantinople; Experienced in His Captivity; and After His Happy Return to His Country, Committed to Writing in the Year of Our Lord 1599.* London: Bell and Daldy, 1862.

Möbius, Friedrich. "Navis Ecclesiae: Sinnschichten des zeitgenössischen Sprachgebrauchs." *Marburger Jahrbuch für Kunstwissenschaft* 22 (1989): 15–22.

Mokyr, Joel. *The Gifts of Athena: Historical Origins of the Knowledge Economy.* Princeton: Princeton University Press, 2002.

Monserrate, Antonio. *The Commentary of Father Monserrate, S.J., on His Journey to the Court of Akbar.* London: Oxford University Press, 1922.

Morris, Sarah P. *Daidalos and the Origins of Greek Art.* Princeton: Princeton University Press, 1992.

Mout, M. E. H. N., and Wim Blockmans. *The World of Emperor Charles V.* Amsterdam: Koninklijke Nederlandse Akademie van Wetenschappen, 2004.

Moxey, Keith P. F. *Peasants, Warriors, and Wives: Popular Imagery in the Reformation.* Chicago: University of Chicago Press, 1989.

Muir, Edward. *Ritual in Early Modern Europe.* Cambridge: Cambridge University Press, 1997.

Müller, Matthias, ed. *Kulturtransfer am Fürstenhof: Hofische Austaushprozesse und ihre Medien im Zeitalter Kaiser Maximilian I.* Berlin: Lukas Verlag für Kunst- und Geistgeschichte, 2013.

Müller, Theodor, and Erich Steingräber. "Die französische Goldemailplastik um 1400." *Münchner Jahrbuch der bildenden Kunst* 5 (1954): 29–79.

Münchner Intelligenzblatt: Churfürstl. Intelligenz- u. Address-Comtoir.

Murphy, Susan. "Heron of Alexandria's *On Automaton Making*." *History of Technology* 17 (1995): 1–44.

Necipoğlu, Gülru. *Architecture, Ceremonial, and Power: The Topkapı Palace in the Fifteenth and Sixteenth Centuries.* New York: Architectural History Foundation, 1991.

———. "Süleyman the Magnificent and the Representation of Power in the Context of Ottoman-Hapsburg-Papal Rivalry." *Art Bulletin* 71, no. 3 (1989): 401–27.

Nelson, Robert S., and Richard Shiff. *Critical Terms for Art History.* Chicago: University of Chicago Press, 2003.

Neue deutsche Biographie. Berlin: Duncker und Humboldt, 1953–2017.

Nicollier-De Weck, Béatrice. *Hubert Languet (1518–1581): Un réseau politique international, de Melanchthon à Guillaume d'Orange.* Geneva: Droz, 1995.

Nonnus. *Dionysiaca.* Translated by W. H. D. Rouse. Cambridge: Harvard University Press, 1940.

Okada, Amina. *Indian Miniatures of the Mughal Court.* New York: Harry N. Abrams, 1992.

Olsen, B. "Material Culture after Text: Remembering Things." *Norwegian Archaeological Review* 36, no. 2 (2003): 87–104.

O'Malley, John W. *The First Jesuits.* Cambridge: Harvard University Press, 1993.

———. *Praise and Blame in Renaissance Rome: Rhetoric, Doctrine, and Reform in the Sacred Orators of the Papal Court, c. 1450–1521.* Durham: Duke University Press, 1979.

Oman, Charles Chichele. *Medieval Silver Nefs.* London: Her Majesty's Stationery Office, 1963.

Orlin, Lena Cowen, and J. Leeds Barroll. *Center or Margin: Revisions of the English Renaissance in Honor of Leeds Barroll.* Selinsgrove, Pa.: Susquehanna University Press, 2006.

Osborne, June. *Entertaining Elizabeth I: The Progresses and Great Houses of Her Time.* London: Bishopsgate Press, 1989.

Pagani, Catherine. *Eastern Magnificence and European Ingenuity: Clocks of Late Imperial China.* Ann Arbor: University of Michigan Press, 2001.

Pal, Pratapaditya. *Master Artists of the Imperial Mughal Court.* Bombay: Marg, 1991.

Parshall, Peter. "Imago Contrafacta: Images and Facts in the Northern Renaissance." *Art History* 16, no. 4 (1993): 554–79.

Pereira, José, and Pratapaditya Pal. *India and Portugal: Cultural Interactions.* Mumbai: Marg, 2001.

Pfister, Louis. *Notices biographiques et bibliographiques sur les Jésuites de l'ancienne mission de Chine, 1552–1773.* Shanghai: Imprimerie de la Mission catholique, 1932.

Pilaski, Katharina. "Der Jesuit im Nierenstein: Einen Merkwürdigkeit aus dem Wittelsbacher Hausarchiv." *Aviso: Zeitschrift für Wissenschaft und Kunst in Bayern* 4 (2005): 20–21.

———. *The Munich Kunstkammer: Art, Nature, and the Representation of Knowledge in Courtly Contexts.* Santa Barbara: University of California, Santa Barbara, 2007.

Pinney, Christopher. "Creole Europe: The Reflection of a Reflection." *Journal of New Zealand Literature,* no. 20 (2002): 125–61.

Pinney, Christopher, and Nicholas Thomas. *Beyond Aesthetics: Art and the Technologies of Enchantment.* Oxford: Berg, 2001.

Polzer, Joseph. "Ambrogio Lorenzetti's 'War and Peace' Murals Revisited: Contributions to the Meaning of the 'Good Government Allegory.'" *Artibus et Historiae* 23, no. 45 (2002): 63–105.

Pope-Hennessy, John Wyndham. *Italian Renaissance Sculpture.* London: Phaidon, 2002.

Preda, Alex. "The Turn to Things: Arguments for a Sociological Theory of Things." *Sociological Quarterly* 40, no. 2 (1999): 347–66.

Press, Volker. "The Habsburg Court as Center of the Imperial Government." *Journal of Modern History* 58 (1986): 23–45.

Price, Derek J. de Solla. "Automata and the Origins of Mechanism and Mechanistic Philosophy." *Technology and Culture* 5, no. 1 (1964): 9–23.

Princely Magnificence: Court Jewels of the Renaissance, 1500–1630. London: Debrett's Peerage in association with the Victoria and Albert Museum, 1980.

Protz, Albert. *Mechanische Musikinstrumente.* Kassel: Bärenreiter, 1939.

Ptak, Roderich, and Dietmar Rothermund. *Emporia, Commodities, and Entrepreneurs in Asian Maritime Trade, c. 1400–1750*. Stuttgart: Steiner, 1991.

Quintilian. *The Orator's Education*. Edited and translated by Donald A. Russell. Cambridge: Harvard University Press, 2001.

Rabelais, François. *Gargantua and Pantagruel*. Penguin Classics. London: Penguin, 2006.

Raby, Julian. *Venice, Dürer, and the Oriental Mode*. London: Islamic Art Publications, 1982.

Ramaswamy, Sumathi. "Conceit of the Globe in Mughal Visual Practice." *Comparative Studies in Society and History* 49, no. 4 (2007): 751–82.

Ramberti, Benedetto. *Libri tre delle cose de Turchi. Nel primo il uiaggio da Venetia à Costantinopoli, con gli nomi de luoghi antichi & moderni: nel secondo la Porta, cioe la corte de soltan Soleymano, signor de Turchi: nel terzo il modo del reggere il stato & imperio suo*. Venice, 1539.

Ramus, Petrus. *Scholarum mathematicarum libri unus et triginta*. Frankfurt, 1627.

Randall, Lilian M. C. *Images in the Margins of Gothic Manuscripts*. Berkeley: University of California Press, 1966.

Rawlinson, H. G. "The Embassy of William Harborne to Constantinople, 1583–8." *Transactions of the Royal Historical Society* 5 (1922): 1–27.

Reber, Franz von. *Die Bildnisse der herzoglich bayerischen Kunstkammer nach dem Fickler'schen Inventar von 1598*. N.p.

Regensburg. *Das Rathaus zu Regensburg: Ein Markstein deutscher Geschichte und deutscher Kunst: Die Stadt Regensburg seit 100 Jahren Bayerns Krone huldigend widmet dieses Werk dem bayerischen Stammlande*. Regensburg: Habbel, 1910.

Reich, Emil. *Select Documents Illustrating Mediaeval and Modern History*. London: P. S. King and Son, 1905.

Reuter, Timothy. "Pre-Gregorian Mentalities." *Journal of Ecclesiastical History* 45, no. 3 (1994): 465–74.

Ricci, Matteo. *China in the Sixteenth Century: The Journals of Matthew Ricci, 1583–1610*. New York: Random House, 1953

Rice, Yael. "The Emperor's Eye and the Painter's Brush: The Rise of the Mughal Court Artist, c. 1546–1627." PhD diss., University of Pennsylvania, 2011.

Riezler, Sigmund von. *Zur Würdigung Herzog Albrechts V. von Bayern und seiner inneren Regierung*. Munich: K. Akademie, 1894.

Riskin, Jessica. "The Defecating Duck; or, the Ambiguous Origins of Artificial Life." *Critical Inquiry* 29, no. 4 (2003): 599–633.

———, ed. *Genesis Redux: Essays in the History and Philosophy of Artificial Life*. Chicago: University of Chicago Press, 2007.

———. "Machines in the Garden." *Republics of Letters: A Journal for the Study of Knowledge, Politics, and the Arts* 1 (2010).

Rizvi, Kishwar. "The Suggestive Portrait of Shah 'Abbas: Prayer and Likeness in a Safavid '*Shahnama*.'" *Art Bulletin* 94, no. 2 (2012): 226–50.

Roby, Courtney Ann. *Technical Ekphrasis in Ancient Science: The Written Machine between Alexandria and Rome*. Cambridge: Cambridge University Press, 2016.

Roe, Thomas. *The Embassy of Sir Thomas Roe to India, 1615–19, as Narrated in His Journal and Correspondence*. London: Oxford University Press, 1926.

Roodenburg, Herman, and Bernd Roeck, eds. *Cultural Exchange in Early Modern Europe*. Cambridge: Cambridge University Press, 2007.

Rosenblum, Robert von, *L'art et les révolutions: Résumés des communications*. Strasbourg: Valblor, 1989.

Rosenthal, Earl E. "The Invention of the Columnar Device of Emperor Charles V at the Court of Burgundy in Flanders in 1516." *Journal of the Warburg and Courtauld Institutes* 36 (1973): 198–230.

———. "Plus ultra, non plus ultra, and the Columnar Device of Emperor Charles V." *Journal of the Warburg and Courtauld Institutes* 34 (1971): 204–28.

———. "Die 'Reichskrone,' die 'Wiener Krone' und die 'Karls des Grossen' um 1520." *Jahrbuch der Kunsthistorischen Sammlungen in Wien* 66 (1970): 7–48.

Rössler, Hans. *Geschichte und Strukturen der evangelischen Bewegung im Bistum Freising 1520–1571*. Nuremberg: Verein für Bayerische Kirchengeschichte, 1966.

Ruh, Kurt, Karl Langosch, Wolfgang Stammler, and Christine Stöllinger-Löser. *Die deutsche Literatur des Mittelalters: Verfasserlexikon*. 14 vols. Berlin: de Gruyter, 1978–2008.

Russo, Alessandra. "A Tale of Two Bodies: On Aesthetic Condensation in the Mexican Colonial Graffiti of Actopan, 1629." *RES: Anthropology and Aesthetics*, nos. 49–50 (2006): 59–79.

Rüstow, Alexander. *Lutherana Tragoedia Artis: Zur geistesgeschichtlichen Einordnung von Hans Baldung Grien angesichts der erstmaligen Ausstellung seines Gesamtwerkes in Karlsruhe*. Zurich: Ges. Schweizer Monatshefte, 1959.

Rutherford, Jonathan, ed. *Identity: Community, Culture, Difference*. London: Lawrence and Wishart, 1998.

Sachs, Hans. *Sämtliche Fabeln und Schwänke*. Halle: M. Niemeyer, 1893.

Santos, João dos. *Ethiopia Oriental, e Varia historia de covsas*

notaueis de Oriente. Évora: Impressa no Conuento de S. Domingos, por Manoel de Lira, 1609.
Schaffer, Simon. "Deus et Machina: Human Nature and Eighteenth Century Automata." *Lettre de la Maison Française d'Oxford* 9 (1998): 9–18.
Schilling, Heinz, ed. *Die reformierte Konfessionalisierung in Deutschland: Das Problem der Zweiten Reformation*. Gütersloh: G. Mohn, 1986.
Schmetzer, Adolf. *Die Restaurierung des Reichssaales und seiner Nebenräume*. Regensburg: Rathaus zu Regensburg, 1910.
Schoch, Rainer, et al., *Albrecht Dürer: Das druckgraphische Werk*. Munich: Prestel, 2001–4.
Schorn-Schütte, Luise. *Karl V: Kaiser zwischen Mittelalter und Neuzeit*. Munich: C. H. Beck, 2000.
Schramm, Percy Ernst. *Herrschaftszeichen und Staatssymbolik: Beiträge zu ihrer Geschichte vom dritten bis zum sechzehnten Jahrhundert*. Stuttgart: Hiersemann, 1954.
———. *Kaiser, Könige und Päpste: Gesammelte Aufsätze zur Geschichte des Mittelalters*. Stuttgart: Hiersemann, 1968.
———. *Die zeitgenössischen Bildnisse Karls des Grossen*. Berlin, 1928.
Schröder, Hans von. *Dirich Utermarke ein hamburger Goldschmied der Renaissance: Von Hans Schröder*. Hamburg: Museum für Hamburgische Geschichte, 1939.
Schulin, Ernst. *Kaiser Karl V: Geschichte eines übergrossen Wirkungsbereiches*. Stuttgart: Kohlhammer, 1999.
Schweigger, Salomon. *Ein newe reyssbeschreibung auss Teutschland nach Constantinopel vnd Jerusalem: darinn die Gelegenheit derselben Länder, Städt, Flecken, Geben &c. der innwohnenten Völcker Art, Sitten, Gebräuch, Trachten, Religion vnd Go Gottesdienst, &c.* . . . Nuremberg, 1619.
Scribner, Robert W. *For the Sake of Simple Folk: Popular Propaganda for the German Reformation*. Cambridge: Cambridge University Press, 1981.
———. "Incombustible Luther: The Image of the Reformer in Early Modern Germany." *Past and Present*, no. 110 (1986): 38–68.
———. "Oral Culture and the Diffusion of Reformation Ideas." *History of European Ideas* 5, no. 3 (1984): 237–56.
———. *Popular Culture and Popular Movements in Reformation Germany*. London: Hambledon, 1987.
Seelig, Lorenz. "Die münchner Kunstkammer: Geschichte, Anlage, Ausstattung." *Jahrbuch der bayerischen Denkmalpflege / hrsg. vom Bayer. Landesamt für Denkmalpflege* 40 (1986): 101–38.
Seelig, Lorenz, and Joaneath Spicer. "Die Gruppe der Diana auf dem Hirsch in der Walters Art Gallery." *Journal of the Walters Art Gallery* 49–50 (1991): 107–18.
Shirley, John W., and F. David Hoeniger. *Science and the Arts in the Renaissance*. Washington, D.C.: Folger Shakespeare Library, 1985.
Sievernich, Gereon, and Hendrik Budde. *Europa und der Orient, 800–1900*. Gütersloh: Bertelsmann, 1989.
Silver, Larry. *Marketing Maximilian: The Visual Ideology of a Holy Roman Emperor*. Princeton: Princeton University Press, 2008.
Silver, Larry, and Elizabeth Wyckoff, eds. *Grand Scale: Monumental Prints in the Age of Dürer and Titian*. New Haven: Yale University Press, 2008.
Singh, Jyotsna G. *A Companion to the Global Renaissance: English Literature and Culture in the Era of Expansion*. Malden, Mass.: Wiley-Blackwell, 2009.
Smith, Jeffrey Chipps, ed. *New Perspectives on the Art of Renaissance Nuremberg: Five Essays*. Austin: University of Texas Press, 1985.
———. *Sensuous Worship: Jesuits and the Art of the Early Catholic Reformation in Germany*. Princeton: Princeton University Press, 2002.
Smith, Jonathan Z. *To Take Place: Toward Theory in Ritual*. Chicago: University of Chicago Press, 1987.
Smyth, Craig, and Gian Carlo Garfagnini. *Florence and Milan: Comparisons and Relations; Acts of Two Conferences at Villa I Tatti in 1982–1984*. Florence: La Nuova Italia editrice, 1989.
Soergel, Philip M. *Wondrous in His Saints: Counter-Reformation Propaganda in Bavaria*. Berkeley: University of California Press, 1993.
Söring, Jürgen, ed. *Androïden: Zur Poetologie der Automaten*. Frankfurt am Main: P. Lang, 1997.
Speel, Erika. *Dictionary of Enamelling*. Aldershot: Ashgate, 1998.
Spence, Jonathan D. *The Memory Palace of Matteo Ricci*. New York: Penguin, 1985.
Spyer, Patricia. *Border Fetishisms: Material Objects in Unstable Spaces*. New York: Routledge, 1998.
St. Clair, Alexandrine N. *The Image of the Turk in Europe*. New York: Metropolitan Museum of Art, 1973.
Stetten, Paul von. *Kunstgewerbe und Handwerksgeschichte der Reichsstadt Augsburg*. Augsburg, 1779.
Stielau, Allison. "The Weight of Plate in Early Modern Inventories and Secularization Lists." *Journal of Art Historiography* 11 (2014): 1–29.
Stockbauer, J. *Die Kunstbestrebungen am bayerischen Hofe unter Herzog Albert V. und seinem Nachfolger Wilhelm V.* Vienna: W. Braumüller, 1874.

Stratton-Pruitt, Suzanne L. *The Virgin, Saints, and Angels: South American Paintings, 1600–1825, from the Thoma Collection*. Stanford: Skira, 2006.

Strauss, Gerald. *Luther's House of Learning: Indoctrination of the Young in the German Reformation*. Baltimore: Johns Hopkins University Press, 1978.

Streng, Friedrich. "Augsburger Meister der Schmiedgasse um 1600." *Blätter des bayerischen Landesverein für Familienkunde* 1 (1963): 260–61.

Strøm-Olsen, Rolf. "Dynastic Ritual and Politics in Early Modern Burgundy: The Baptism of Charles V." *Past and Present*, no. 175 (2002): 34–64.

Subrahmanyam, Sanjay. *Beyond Incommensurability: Understanding Inter-imperial Dynamics*. Theory and Research in Comparative Social Analysis 32. Los Angeles: UCLA Department of Sociology, 2005.

———. *Courtly Encounters: Translating Courtliness and Violence in Early Modern Eurasia*. Cambridge: Harvard University Press, 2012.

———. *The Portuguese Empire in Asia, 1500–1700: A Political and Economic History*. London: Longman, 1993.

———. "A Roomful of Mirrors: The Artful Embrace of Mughals and Franks, 1550–1700." *Ars Orientalis* 39 (2010): 39–83.

———. *Three Ways to Be Alien: Travails and Encounters in the Early Modern World*. Waltham: Brandeis University Press, 2011.

Swan, Claudia. "Ad Vivum, Naer Het Leven, from the Life: Defining a Mode of Representation." *Word and Image* 11, no. 4 (1995): 353–72.

———. *Art, Science, and Witchcraft in Early Modern Holland: Jacques de Gheyn II (1565–1629)*. Cambridge: Cambridge University Press, 2005.

Syndram, Dirk. *The Green Vault in Dresden: Renaissance and Baroque Treasury Art*. Munich: Deutscher Kunstverlag, 2005.

———. "Oral Culture and the Diffusion of Reformation Ideas." *History of European Ideas* 5, no. 3 (1984): 237–56.

. *Renaissance and Baroque Treasury Art: The Green Vault in Dresden*. Dresden: Staatliche Kunstsammlungen Dresden, 2004.

———. *Schatzkunst der Renaissance und des Barock*. Munich: Kunstverlag, 2004.

Syndram, Dirk, Martina Minning, and Jochen Vötsch. *Die Inventare der kurfürstlich-sächsischen Kunstkammer in Dresden*. Dresden: Michel Sandstein, 2010.

Syndram, Dirk, and Antje Scherner, eds. *Princely Splendor: The Dresden Court, 1580–1620*. New York: Metropolitan Museum of Art, 2004.

Tanner, Marie. *The Last Descendant of Aeneas: The Hapsburgs and the Mythic Image of the Emperor*. New Haven: Yale University Press, 1993.

Tanner, Norman, and Sethina Watson. "Least of the Laity: The Minimum Requirements for a Medieval Christian." *Journal of Medieval History* 32, no. 4 (2006): 395–423.

Terry, Edward. *A Voyage to East-India; Wherein Some Things Are Taken Notice Of, in Our Passage Thither, but Many More in Our Abode There, within That Rich and Most Spacious Empire of the Great Mogul: Mixt with Some Parallel Observations and Inferences upon the Story, to Profit as Well as Delight the Reader*. London: J. Wilkie, 1777.

Terzioğlu, Derin. "The Imperial Circumcision Festival of 1582: An Interpretation." *Muqarnas* 12 (1995): 84–100.

Thiessen, Gesa Elsbeth, ed. *Theological Aesthetics: A Reader*. London: S.C.M. Press, 2004.

Thomas, Andrew L. *A House Divided: Wittelsbach Confessional Court Cultures in the Holy Roman Empire, c. 1550–1650*. Leiden: E. J. Brill, 2010.

Tietze, Hans. *Die Denkmale des politischen Bezirkes Melk*. Vienna: A. Schroll, 1909.

Tipton, Susan. *Res publica bene ordinata: Regentenspiegel und Bilder vom guten Regiment: Rathausdekorationen in der frühen Neuzeit*. Hildesheim: G. Olms, 1996.

Trexler, Richard C. *The Journey of the Magi: Meanings in History of a Christian Story*. Princeton: Princeton University Press, 1997.

Trnek, Helmut, and Nuno Vassallo e Silva, eds. *Exotica: The Portuguese Discoveries and the Renaissance Kunstkammer*. Lisbon: Calouste Gulbenkian Foundation, 2001.

Truitt, Elly Rachel. *Medieval Robots: Mechanism, Magic, Nature, and Art*. Philadelphia: University of Pennsylvania Press, 2015.

Turner, Victor. *Dramas, Fields, and Metaphors: Symbolic Action in Human Society*. Ithaca: Cornell University Press, 1974.

Tybjerg, Karin. "Wonder-Making and Philosophical Wonder in Hero of Alexandria." *Studies in History and Philosophy of Science* 34 (2003): 443–66.

Vantuono, William. *Sir Gawain and the Green Knight: A Dual-Language Version*. New York: Garland, 1991.

Vasari, Giorgio. *The Lives of the Artists*. Translated by Julia Conaway Bondanella and Peter Bondanella. Oxford: Oxford University Press, 1998.

Vitruvius Pollio. *De Architectura*. Translated by Frank Stephen Granger. Cambridge: Harvard University Press, 1955.

Voskuhl, Adelheid. *Androids in the Enlightenment: Mechanics, Artisans, and Cultures of the Self.* Chicago: University of Chicago Press, 2013.

Warncke, Carsten-Peter. "Cellinis 'Saliera': Der Triumph des Goldschmieds." *Artibus et Historiae* 21, no. 42 (2000): 41–52.

Warner, David A. "Ritual and Memory in the Ottonian Reich: The Ceremony of Adventus." *Speculum* 76, no. 2 (2001): 255–83.

———. "Thietmar of Merseburg on Rituals of Kingship." *Viator* 26 (1995): 53–76.

Watanabe-O'Kelly, Helen. *Court Culture in Dresden: From Renaissance to Baroque.* New York: Palgrave, 2002.

Weitzman, Steven. *Solomon: The Lure of Wisdom.* New Haven: Yale University Press, 2011.

Wentersdorf, Karl P. "The Symbolic Significance of Figurae Scatologicae in Gothic Manuscripts." In *Word, Picture, and Spectacle*, edited by Clifford Davidson, 1–19. Kalamazoo: Medieval Institute, Western Michigan University, 1984.

Westfehling, Uwe. *Die Messe Gregors des Grossen: Vision, Kunst, Realität: Katalog und Führer zu einer Ausstellung im Schnütgen-Museum der Stadt Köln.* Cologne: Schnütgen Museum, 1982.

Willigen, Adriaan van der. *Les artistes de Harlem: Notices historiques avec un précis sur la guilde de St. Luc.* Nieuwkoop: B. de Graaf, 1970.

Wilson, Bronwen. "Reflecting on the Turk in Late Sixteenth-Century Venetian Portrait Books." *Word and Image* 19, nos. 1–2 (2003): 38–58.

———. *The World in Venice: Print, the City, and Early Modern Identity.* Toronto: University of Toronto Press, 2005.

Wilson, Jean C. *Painting in Bruges at the Close of the Middle Ages: Studies in Society and Visual Culture.* University Park: Pennsylvania State University Press, 1998.

Wink, André. *Akbar.* Oxford: Oneworld, 2009.

Wood, Christopher S. *Forgery, Replica, Fiction: Temporalities of German Renaissance Art.* Chicago: University of Chicago Press, 2008.

Wright, Elaine Julia. *Muraqqa': Imperial Mughal Albums from the Chester Beatty Library, Dublin.* Alexandria, Va.: Art Services International, 2008.

Wunder, Amanda. "Western Travelers, Eastern Antiquities, and the Image of the Turk in Early Modern Europe." *Journal of Early Modern History* 7, no. 1 (2003): 89–119.

Yates, Frances A. *Astraea: The Imperial Theme in the Sixteenth Century.* London: Routledge, 1975.

Yurdusev, A. Nuri, ed. *Ottoman Diplomacy: Conventional or Unconventional?* New York: Palgrave Macmillan, 2004.

Županov, Ines G. *Missionary Tropics: The Catholic Frontier in India, 16th–17th Centuries.* Ann Arbor: University of Michigan Press, 2005.

Index